Life Cycle Assessment in the Chemical Product Chain

Simone Maranghi • Carlo Brondi
Editors

Life Cycle Assessment in the Chemical Product Chain

Challenges, Methodological Approaches and Applications

Springer

Editors
Simone Maranghi
Biotechnology, Chemistry and Pharmacy
University of Siena
Siena, Italy

Carlo Brondi
STIIMA-CNR
Milan, Italy

ISBN 978-3-030-34423-8 ISBN 978-3-030-34424-5 (eBook)
https://doi.org/10.1007/978-3-030-34424-5

This Springer imprint is published by the registered company Springer Nature Switzerland AG
The registered company address is: Gewerbestrasse 11, 6330 Cham, Switzerland

Foreword

The chemical industry has a crucial role in the development of a competitive and sustainable future. Both industrial, political and scientific roadmaps put the emphasis on transforming such sector by a sustainability paradigm. The recent ambitious EU energy and climate plan, as well as Agenda for Sustainable Development, still underline the needs of a radical change in the world economy by its shift towards less impacting industrial sectors.

Within the process industry, the chemical industry has a relevant effect on product chains. From one side, the chemical industry is an energy-demanding and emission-producing sector. On the other hand, its products have a double effect. Chemicals are crucial both to enable the energy saving, to improve energy efficiency and to reduce greenhouse gas emissions in economic sectors as well as in producing specific types of harmful waste and emission along the consumer product chain.

Sustainability-related decisions in the field of chemical products require science-based approaches, which focus on process design and development, and the analysis of processing chains. The development of innovative future solutions strictly needs to be coupled to adequate methodologies in order to prove its effective impact at the global level.

In such a context, Life Cycle Assessment (LCA) methodology is pivotal for this purpose. It is a well-known scientific methodology to assess the energy and environmental impacts of products, processes and activities, encompassing extraction and processing of raw materials, manufacturing and assembly processes, product distribution, use, maintenance and end use. Thus, it is useful to avoid burden transfer from one life cycle step to another and from an impact category to the others. Furthermore, it helps in supporting the identification of priority actions in policymaking, the selection of the best low-carbon solutions for resources supply and use, the identification of the hotspots for reducing the carbon intensity of chemical products and the management of the end-of-life of such products.

However, modelling of chemical processes' impacts has to face methodological barriers, due to the complexity and specificity of the chemical product chain, and to the lack of specific methodologies to transform field data into reliable data for comparison.

In this context, this book aims to share some methodological researches on LCA applied to the chemical products and processes and provides an overview of the most recent outcomes, approaches and tools for modelling chemicals in an LCA perspective. Each chapter will address specific problems in modelling or specific issues in LCA application to relevant cases.

In detail, the book consists of two parts. The first one addresses general modelling issues, dealing with general issues and methods for modelling, an adaptation of field data to other existing database, life cycle inventory procedure, allocation rules, selection of suitable indicators for Life Cycle Impact Assessment (LCIA) and integration of LCIA indicators with existing sustainability indicators in the industry sector.

The second part of the book presents the LCA application to specific field contexts, concerning the chemical processes at different scales or areas. In addition to the general highlighted outcomes from the first part, other barriers related to the application context are explored, and the application of the LCA method in a circular economy perspective of the chemical industry is discussed.

The book is the outcome of the activity of several members of the working group 'Chemical Processes and Products' of the 'Italian LCA Network' association. I would like to acknowledge all authors for submitting their valuable work and the members of the 'Italian LCA Network' association for supporting this work.

This book can represent for LCA-practitioners, researchers and students an opportunity to learn more applications of the LCA methodology, and to understand the environmental impact of chemical processes and products, through the assessment of their life cycles.

The Italian LCA Network

The Italian LCA Network was created in 2006 with the aim to have a network for exchanging information, methodologies and good practices on LCA in Italy. In 2012, the Italian LCA Network became an association, founded by the Italian National Agency for New Technologies, Energy and Sustainable Economic Development (ENEA); the Politecnico of Milano; the Universities of Bari, Chieti-Pescara, Padova and Palermo and the National Interuniversity Consortium for Chemical Reactivity and Catalysis (CIRCC).

The main objectives of the 'Italian LCA Network' association are the following: promoting the adoption of the life cycle thinking approach for achieving a sustainable development; promoting the dissemination of the LCA methodology at national level and the exchange of information and best practices on the LCA in Italy and encouraging networking processes among different stakeholders for the realisation of national and international projects.

The Working Group 'Chemical Processes and Products'

The main goals of the working group 'Chemical processes and products' of the Italian LCA Network are the assessment of the environmental sustainability of chemical processes and products, making in contact many research groups that are involved in the application of LCA at several chemical sectors: Green Chemistry, Green Metrics, industrial processes, pharmaceutical industries and laboratory researches.

Palermo, Italy

Maurizio Cellura
The President of the Italian LCA
Network from 2015–2019

Preface

Each chapter of this book addresses specific problems in modelling or specific issues in Life Cycle Assessment (LCA) application to chemical sector context. In particular, aims to support analyst in the modelling perspective. In fact, the first part, *Transversal Modelling Issues in LCA for Chemicals*, addresses specific modelling issues within the chemical sector. Modelling issues are related to specific LCA maturity level for its application to such sector and include topics such as chemicals' modelling according to limited database availability or chemicals' modelling by considering circular economy aspects. The second part, *Application of LCA in Chemicals Process Modelling*, deals with LCA application to specific application contexts. Such parts are referred to the subject of the LCA modelling and provide specific suggestion in order to model specific areas from laboratory scale up to pilot scale.

The final outcome of the book is intended to support in a first stage chemical experts and LCA analysts in properly assessing the environmental effects of chemicals. Such need emerges from improved standard for sustainability in the advanced industry as well as in a further progress and standardisation of the LCA procedures at European and global level. The presented issues are not exhaustive; nevertheless, the book aims to place a milestone in the definition of chemical sector issues for the application of LCA.

Part I—Transversal Modelling Issues in LCA for Chemicals

Modelling of environmental and social impacts of chemical processes meets many methodological barriers. These barriers are linked to the complexity and specificity of the chemical product chain and to the lack of specific methodologies to transform field data into reliable results for comparison. The 'Part I' of such book aims to analyse the general issues for the LCA methodology in the chemical sector.

Chapter 1 faces the problems of the Life Cycle Inventory (LCI) analysis for chemical processes and the lack of data concerning inventories of basic and fine chemical substances. The chapter aims to illustrate how the scientific LCA community has proposed to overcome the lack of data concerning inventories of basic and fine chemicals both for input and output flows, in order to investigate the main advantages and drawbacks of each approach.

Chapter 2 reviews the different impact assessment methods used in the LCA of chemicals and identifies the main categories of environmental impact assessed using these methods. A paragraph of the chapter is devoted to human- and eco-toxicity, which are difficult to treat within the life cycle impact assessment because they reflect local impacts rather than global ones.

Chapter 3 addresses the integration of LCA to sustainable principles and tools in the Green Chemistry sector, discussing the integration of LCA approach with the most consolidated tools and methodologies. Also, the chapter discloses how the chemical industries deal with the issue of social LCA and how this kind of assessment is communicated to consumers, by presenting both the most consolidated approaches and the simplest ways to perform such evaluation within the chemical sector.

Chapter 4 deals with different case studies of industrial symbiosis and how these are treated according to literature, regulatory approaches and main guidelines. Once the different approaches and their range of application are presented, the chapter describes the main barriers and strengths in the application of these rules to the chemical sector. Finally, the identified approaches are applied to a specific case of industrial symbiosis in the chemical sector by illustrating how the single data can be calculated and can quantitatively change.

Part II—Application of LCA in Chemicals Process Modelling

The 'Part II' of the book focuses on adapting the LCA methodology to practical contexts concerning the chemical processes at different scales or areas. In addition to the general highlighted outcomes from the first part, other additional barriers that are related to the application context will be explored.

Chapter 5 addresses the LCA integration as a decisional tool in the early design stage at laboratory scale with other current green metric parameters and tools, emphasising main advantages and limitations. The importance to integrate the LCA methodology with further green metrics and evaluation tools at a laboratory scale relies on the fact that very often more affordable and straightforward methods can be used to obtain a trustworthy environmental evaluation, which very often is in full agreement with the LCA results.

Chapter 6 deals with the problem of LCA modelling during the process scale-up from laboratory prototypal process up to industrial production. The purpose is to review the different approaches and methodologies reported in the literature, highlighting what the main critical issues that need to be faced with the application of LCA to systems characterised by a different production scale are. The application of different scale-up approaches, such as mathematical simulations, extrapolation functions, statistical models and physical–chemical approaches, is reviewed.

Chapter 7 addresses the application of LCA as a tool for the environmental assessment of a system in the context of a circular economy for the chemical industry. In particular, the LCA methodology allows for a comprehensive evaluation in order to highlight how to tackle chemical processes that use waste as secondary raw materials. A specific case study concerning the application of LCA to a real industrial process is discussed, to emphasise potentialities and contextualise uncertainties in LCA modelling for the circular economy.

Siena, Italy Simone Maranghi
Milan, Italy Carlo Brondi

Contents

Part I
Transversal Modelling Issues in LCA for Chemicals

1 The Availability of Suitable Datasets for the LCA Analysis of Chemical Substances

Serena Righi, Alessandro Dal Pozzo, Alessandro Tugnoli, Andrea Raggi, Beatrice Salieri and Roland Hischier

Abstract

The practicality of adopting Life Cycle Assessment to support decision-making can be limited by the resource-intensive nature of data collection and Life Cycle Inventory modelling. The number of chemical products increases continuously, and long-term analyses show that overall growth of chemicals production and demand as well as faster growth in emerging regions is a behaviour that is expected to continue in the future. Regrettably, chemical inventories are typically among the most challenging to model because of the lack of available data and the large number of material and energy flows. This explains why it is so important for the Life Cycle Assessment community to have effective methods to implement life cycle inventories of chemicals available. This chapter deals with the issues of Life Cycle Inventory analysis for chemical processes and the related lack of data concerning inventories of basic and fine chemicals substances. The overall aim of the chapter is to illustrate the different possibilities/approaches that the scientific Life Cycle Assessment community

S. Righi (✉)
DIFA Department of Physics and Astronomy and CIRSA Inter-Departmental Centre for Research in Environmental Sciences, University of Bologna, Bologna, Italy
e-mail: serena.righi2@unibo.it

A. Dal Pozzo · A. Tugnoli
Department of Civil, Chemical, Environmental, and Materials Engineering, University of Bologna, Bologna, Italy

A. Raggi
Department of Economic Studies, University "G. d'Annunzio" of Chieti-Pescara, Pescara, Italy

B. Salieri · R. Hischier
Swiss Federal Laboratories for Materials Science and Technology (Empa), St. Gallen, Switzerland

S. Maranghi and C. Brondi (eds.), *Life Cycle Assessment in the Chemical Product Chain*, https://doi.org/10.1007/978-3-030-34424-5_1

has developed in order to overcome such a lack of data concerning the inventories of a specific (basic and/or fine) chemical substance both for input and output flows. Their main advantages and drawbacks are identified and discussed briefly.

1.1 Introduction

The Life Cycle Inventory (LCI) analysis step within the ISO 14040 and 14044 standards involves the compilation and quantification of input/output data for a given product system throughout its life cycle (ISO 2006a, b). Data concerning energy and raw material inputs, products and co-products, waste, emissions to air, discharges to water and soil and other environmental aspects have to be collected (ISO 2006a, b). LCI is considered the most time-consuming, complicated and resource-demanding part of a Life Cycle Assessment (LCA) study (Laurent et al. 2014) and therefore a very critical phase within the entire LCA activities, as the degree of quantification of the inputs and outputs directly affect the following impact assessment and interpretation activities.

This challenge is particularly true in the case of chemical products. Tens of thousands of chemicals are currently in commerce, and hundreds more are introduced every year (USEPA 2016a). In the United States of America, the Toxic Substances Control Act (TSCA) Inventory listed about 85,000 chemicals in 2016 (USEPA 2016b). In Europe, over the first 10 years of the REACH Regulation, nearly 90,000 registrations for chemicals manufactured in, or imported to the EU at above one tonne a year have been submitted (ECHA 2018). The long-term analysis shows that overall growth of chemicals production and demand as well as faster growth in emerging countries is a trend that should continue in the near future. World chemicals sales are expected to reach the level of €6.3 trillion in 2030 (Cefic 2018). To compound the problem, for many of the fine chemicals the bill of materials may involve anywhere from twenty, fifty or more chemical compounds (depending on the complexity), each of which will require their own inventory data to accomplish the assessment (Jiménez-González and Overcash 2014). Because there are so many chemicals—and establishing respective data inventories is expensive and time-consuming—only a small number of today's chemicals are represented in current LCI databases.

Compiling LCIs of chemical compound production can be a complex and challenging endeavour. However, several methods have been proposed in recent decades to facilitate their creation. This chapter aims to illustrate how the scientific LCA community has proposed to overcome the lack of data concerning life cycle inventories of chemicals and investigates the main advantages and drawbacks of each of these approaches.

1.2 Life Cycle Inventory Approaches for Chemicals

In the past decades, several efforts have been made to categorize the methods applied for LCI compilation. Suh and Huppes (2005) identified three different types of approaches, i.e. computational approaches, the economic input–output (EIO) analysis, and combinations of those two approaches. Being part of the first type, process flow diagrams are the oldest and the most common practice in LCI compilation, showing how processes of a product system are interconnected through commodity flows. Each process is represented as a ratio between several inputs and outputs. Using plain algebra, the amount of commodities fulfilling a certain functional unit is obtained. The second computational approach reviewed by Suh and Huppes (2005) is the matrix representation where a system of linear equations is used to solve an inventory problem. Next to this, the authors examined the application of EIO within LCA—starting in early 1990s, when macroeconomic models are combined with sector-level environmental data to estimate total supply-chain impacts of the production (Moriguchi et al. 1993). Hybrid approaches —linking process-based and EIO-based analysis and attempting to exploit the respective strengths and advantages of each of these two approaches—can be further distinguished into the following types: tiered hybrid analysis; IO-based hybrid analysis; integrated hybrid analysis. This review by Suh and Huppes (2005) is not focused on LCA applied on chemistry but it is the first clear classification of LCI methods and cites several case studies related to synthetic (chemical) products (such as e.g. Joshi 1999; Strømman 2001).

In 2014, Jiménez-González and Overcash examined the evolution of the application of LCA in the pharmaceutical and the chemical sector and analysed various methods for gathering inventory data (Jiménez-González and Overcash 2014). They fundamentally retraced the categories identified by Suh and Huppes (2005), thus listing process-based inventories, economic input–output inventories and hybrid approaches. In addition, they included methods such as industrial groups sharing manufacturing data with a third party (Boustead 2005) or streamlined tools (such as Wernet et al. 2009).

In their work about uncertainties within LCI data, Williams et al. (2009) introduced a different nomenclature scheme, distinguishing between bottom-up, top-down and hybrid approaches. In a bottom-up approach, each process along the supply chain of a product is described in terms of its proper inputs and outputs. Top-down approaches start from general, often economic and/or economy-based, data with the aim to extract process-specific information in the form of economic input–output inventories and hybrid approaches are a combination of the two approaches. Such a distinction between bottom-up and top-down approaches was proposed in recent years in several articles concerning new methods with a high aggregated level for implementing LCI of chemical products, like Cashman et al. (2016), Mittal et al. (2018), or de Camargo et al. (2018).

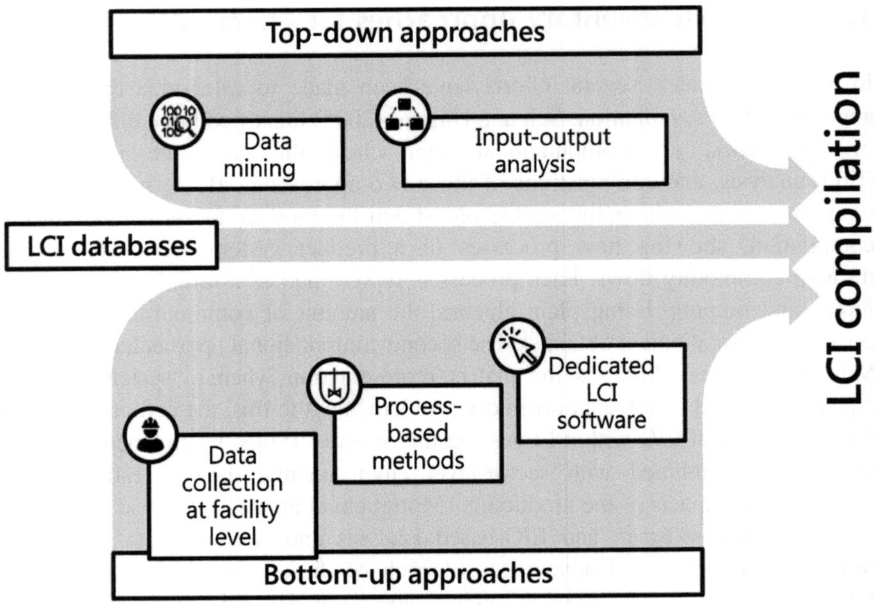

Fig. 1.1 Approaches to the generation of LCI data

In this chapter, this classification in bottom-up, top-down and hybrid approaches is followed in order to guarantee a broader classification and thus a more comprehensive coverage of the issue of finding suitable LCI data for chemicals. Figure 1.1 summarizes the different methods covered in the chapter.

1.3 Bottom-Up Approaches

Bottom-up approaches move from the most detailed level towards the most general. In LCI, a bottom-up approach starts from a finely granulated, detailed process system by identifying the smallest transformation activities (covering the various inputs and outputs as shown in Fig. 1.2) and studying them attentively. These activities are then combined to form larger processes, with a successive incorporation of smaller processes into larger processes, until the entire system is implemented.

1.3.1 Direct Data Collection and Existing LCI Databases

Outlined as above, the LCI compilation from a bottom-up standpoint requires the collection of quantitative information about inputs and outputs, unit process after unit process.

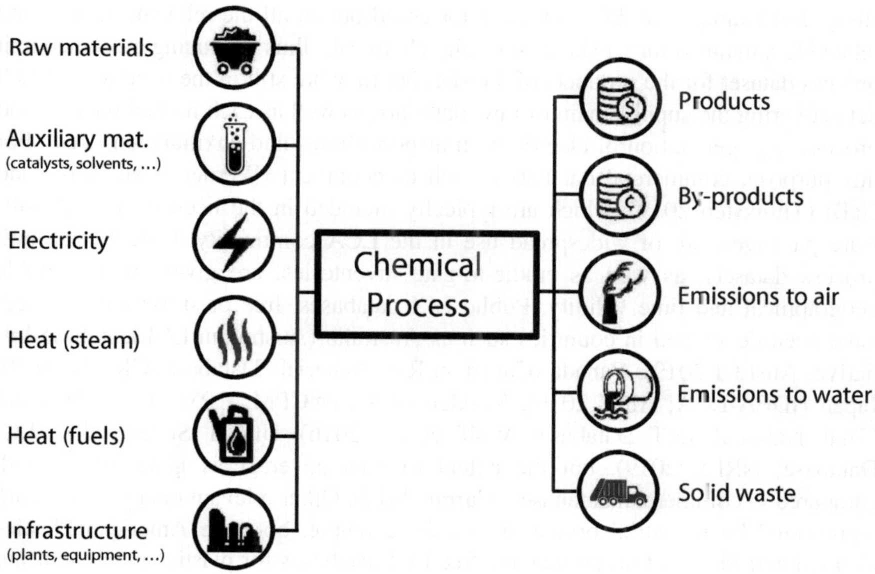

Fig. 1.2 Inventory associated with a chemical process

Ideally, data collection for the unit process inventories is done at a relevant production/manufacturing site (primary data). This means gathering raw data at the plant and transforming them into inventory entries (material and energy inputs, co-products, emissions, waste streams). Raw data at facility level can be obtained from several relevant sources, e.g. by consulting bill of materials, collecting process monitoring data, proposing questionnaires or surveys to the plant personnel and performing on-site measurements. Conducting a proper and meaningful collection of raw data at plant for LCI purposes is not an easy task. Apart from the considerable effort in gathering all the relevant information, the LCA analyst has to reconcile discrepancies between different data sources in the plant and properly validate the resulting dataset by means of consistency and completeness checks. A structured approach is highly recommended and standards (ISO 14040 and 14044) offer useful guidance to the generation of unit process datasets from primary data. Further suggestions are given by the *Global Guidance Principles for Life Cycle Assessment Databases*, a publication issued under the auspices of the United Nations Environment Programme (Sonnemann et al. 2013).

An evident hindrance to the collection of relevant primary data for industrial operations at a chemical processing plant is represented by confidentiality issues. Frequently, information cannot be disclosed by companies or the time required to get publication clearance is not compatible with the time constraints of the LCA study.

Even when access to primary data at the plant is possible, data other than first-hand plant information (secondary data) are still needed to complete the LCA study. For example, an LCA investigator could obtain all the relevant information related to a manufacturing site of specialty chemicals, thus producing a detailed unit process dataset for the products of the site, but they are still in the need to find LCI data covering the supply chain of raw materials, as well as each related background process (e.g. generation of electricity, transport phases and auxiliary services). For this purpose, commercial databases, such as ecoinvent (Wernet et al. 2016) and GaBi (Thinkstep 2019), which are typically included in the licence of LCA software packages, are of widespread use in the LCA community. Gate-to-gate unit process datasets, as well as cradle-to-gate inventories, are given, with variable geographical and time validity. Public LCI databases, free or subscription-based, have been developed in countries such as Australia (Australian LCI Database Initiative; AusLCI 2019), Canada (Canadian Raw Materials Database; CRMD 2019), Japan (IDEA-LCA; AIST 2019), Sweden (SPINE@CPM; CPM 2019), Thailand (Thai National LCI Database; Wolf et al. 2016) United States (U.S. LCI Database; NREL 2019), but the extent of data coverage is generally limited, compared to commercial databases (Curran 2012). Other relevant data providers are represented by industrial organizations: for example, both the American Plastics Council and Plastics Europe manage free LCI databases for plastics manufacturing created on information provided by member companies.

At the European level, a harmonization effort of LCI databases has been started with the creation of the Life Cycle Data Network (JRC 2019), a common infrastructure where data from different organizations are published upon compliance with entry-level requirements. In the framework of the *Single Market for Green Products* initiative launched in April 2013, the European Commission proposed the development of the so-called Product Environmental Footprint (PEF) as a common method of measuring the environmental performance of products, with the aim to standardize the communication of the environmental impacts of products from companies to consumers. For several product categories, PEF category rules (PEFCRs) have been developed to provide specific guidance in LCI and LCIA calculations, with the ultimate aim of bolstering reproducibility and comparability of LCA results. The PEFCRs present the guidelines that companies should follow in the calculation of the PEF of their products and explicate how to refer to secondary datasets for all the raw, intermediate and auxiliary materials that are not produced by the company. PEF-compliant secondary datasets are already in development and are distributed through the Life Cycle Data Network. With particular reference to the use of chemicals, in the PEFCRs developed during the pilot stage of the initiative significant effort has been devoted to associate specific functions required in the manufacturing of a product to the families of chemical compounds that can have that function and to identify reference substances within each family for which LCI data are available. For example, in the production of leather, a required auxiliary compound is the tanning agent. The PEFCR for leather (De Rosa-Giglio et al. 2018) identifies categories of chemical products suitable for tanning (mineral agents, synthetic organic agents and vegetable tannins) and families within the categories (e.g. Al-, Cr-, or Zr-based agents in the mineral agents category), for which at least a representative compound presents a related LCI dataset available (e.g. aluminium sulphate for the Al-based agents). This systematization effort, which will be replicated in all the future PEFCRs, has the goal to simplify and homogenize the use of secondary datasets when primary data are lacking.

1.3.2 Process-Based Methods

Process-based methods of different levels of complexity focus on the production process to obtain the inventory data related to the production of a given chemical substance. Available approaches are vastly different in terms of data/time requirements and resulting accuracy (Parvatker and Eckelman 2019). Here, process-based methods are classified into three categories, depending on the information required for their use and the kind of LCI data that can be extracted (see Table 1.1): process chemistry, conceptual process design, and process modelling and simulation.

1.3.2.1 Process Chemistry

In the absence of any process data, the basis for estimating the LCI associated to the production of a given chemical substance is the reaction stoichiometry. For known chemicals, published balanced reaction pathways can be found easily in open datasets, such as the *Ullmann's Encyclopedia of Industrial Chemistry* (Elvers 2011) or the *Kirk-Othmer's Encyclopedia of Chemical Technology* (Kroschwitz and Seidel 2004), while for a newly synthesized product or a novel synthesis route developed at laboratory scale, stoichiometric data are directly available to the investigator from her own experimental protocol. It should come to no surprise that inventory compilation based on stoichiometry constitutes probably the most common approach for LCI data collection and even established LCI databases rely heavily on stoichiometric assumptions (Hischier et al. 2005).

Table 1.1 Hierarchy of process-based methods for generating LCI data

Approach	Information available/calculation required	LCI generation	
		Material flows	Energy flows
Process chemistry	Reaction stoichiometry, yields, heat of reaction	Reactants, products	Reaction-related energy demand
Conceptual process design	Unit operations, mass and energy balances	Reactants, products, auxiliary materials, waste flows, fugitive emissions (*rules of thumb estimate*)	Process-related energy demand, losses (*rules of thumb estimate*)
Process modelling and simulation	Equipment sizing, utilities, possible energy integration networks	Reactants, products, auxiliary materials, waste flows, fugitive emissions	Plant-related energy demand, losses

Reaction stoichiometry allows calculating, for a unit of given chemical produced, the associated mass flows of reactants, by-products (emissions for LCI purposes) and co-products (if present, they are used for allocation or system expansion). If known, yield of reaction can be taken into account in the calculation. Regarding energy flows, basic thermodynamic data of the involved chemical species (heat of formation and heat capacity) can be used for a raw estimate of the energy requirements for heating (endothermic reactions) or cooling (exothermic reactions). The heat associated with a chemical reaction is the algebraic sum of the heats of the formation of reactants and products. The heat of formation of a compound can be retrieved from databases, e.g. the *NIST-JANAF Thermochemical Tables* (Chase 1998). In the absence of data, expeditious methods such as the group additivity approach (Holmes and Aubry 2012), which requires only the knowledge of the chemical structure of the compound, ensure an acceptable estimate of the heat of formation for LCA purposes. Likewise, the sensible heat to be provided to reactants to reach reaction conditions can be calculated from the specific heat capacity of the compounds, which can be retrieved from sources like the *NIST Chemistry WebBook* (Linstrom and Mallard 2019).

There are clear limitations to the information that can be extracted from reaction stoichiometry. This approach allows the main raw materials and products of the synthesis route to be traced, but neglects auxiliary materials, such as catalysts and solvents. Both classes of materials cannot be excluded without due consideration. As for solvents, e.g. 80–90% of the total mass involved in the production of an active pharmaceutical ingredient can be ascribed to solvent use (Raymond et al. 2010). Furthermore, the stoichiometric approach is inherently limited to reactions and ignores pre- and post-reaction stages. In particular, it excludes from LCI compilation the separation and purification stages, which in several contexts like, e.g. the bio-based industry (Huang et al. 2008), are the most energy-intensive steps.

1.3.2.2 Conceptual Process Design and Scale-Up Methods

To increase the level of detail and the representativeness of the estimated LCI, it is necessary to move up from process chemistry to process design considerations. The definition of the process requires: listing all direct and indirect chemicals involved in the process, determining the conditions for each operation (temperature, pressure and composition of process streams), assuming the efficiencies in chemical conversion and separation stages, and, finally, elaborating the flow diagram of the process (Jimenez-Gonzalez et al. 2000). As a result, the definition of a conceptual process flowsheet and the related unit operations allows drawing the mass and energy balances that provide the material and energy flows of the process.

The data needed to define the process can be extracted from relevant scientific or technical literature and from patent information. When it is not possible to retrieve adequate process details, information gaps can be filled by reasonable estimates and assumptions based on chemical engineering knowledge. Unfortunately, there is no general protocol for process flowsheeting (Papadokonstantakis et al. 2016). Nonetheless, general good practices for early process design, ranging from simple heuristics to short-cut models, have been developed and consolidated in the discipline. Guidance in the selection of assumptions (process yield, solvent recycling, heat recovery efficiency, etc.), as well as empirical design equations, can be found in reference texts such as the *Coulson and Richardson's Chemical Engineering Series* (Sinnott 2005) or the *Perry's Chemical Engineers' Handbook* (Green and Perry 2007). Losses both in terms of energy and materials can be estimated following rules of thumb (Hall 2017). Diffuse and fugitive emissions can also be approximated using generic emission factors, like the ones collected by the US EPA under the *AP-42 Compilation of Air Pollutant Emission Factors* (US EPA 2010). Simplified methods to evaluate storage emissions or process vent emissions are proposed by Smith et al. (2017).

In addition, if the process under study can be decomposed into subprocesses, it is worth recalling that the LCA analyst can rely on established LCI databases (ecoinvent, Thinkstep, etc.) to model rather 'standard' process blocks and perform conceptual process design just for the most case-specific operations, filling specific gate-to-gate data gaps.

In many cases, the LCA practitioner dealing with process design considerations needs to analyse the environmental footprint of a novel chemical or a new route for the synthesis of a known chemical, discovered at the laboratory stage. Since the main interest is to develop a credible LCI of the theoretical future industrial scale, rather than conduct the mere LCA of the laboratory-based process, the investigator has to perform a prospective scale-up. The primary data gathered at laboratory level (material/energy inputs) cannot be generally considered representative of the same synthesis performed at commercial scale (Khoo et al. 2018). At lab stage, the aim of the researchers is just to demonstrate that a certain synthesis protocol works and the process itself is not optimized in terms of consumption of energy and auxiliary materials (Hischier et al. 2018). Several impacts can be reduced significantly when the process is established at industrial scale, thanks to an increased technological maturity, the advantages of economies of scale, and opportunities for heat integration and materials recycling (Arvidsson and Molander 2016; Gavankar et al. 2015).

A full process scale-up procedure would require a detailed feasibility study, generally involving fluid-dynamic modelling of equipment, pilot-scale experimental campaigns and process optimization trials (Righi et al. 2018). Nonetheless, simplified approaches have been proposed for the necessities of scale-up in the context of LCI data generation. This issue will be dealt in detail in Chap. 6. Here, it is just mentioned that process scale-up for LCA purposes is a topic of vivid discussion in the recent scholarly literature. Several considerations are based on expert judgment, e.g. in assessing how much solvent could be spared in industrial operation compared to the laboratory synthesis of a novel chemical compound (Hischier et al. 2018). In analogy with well-known scaling rules historically proposed for equipment cost estimation, other authors suggested the use of power-law relationships to estimate, e.g. how energy consumption scales with process size (Caduff et al. 2014). A more systematic approach is represented by scale-up frameworks proposed, e.g. by Shibasaki et al. (2007) and Zhou et al. (2017), with specific reference to the transition between pilot scale and commercial scale and, e.g. by Piccinno et al.

(2016) and Simon et al. (2016), concerning the scaling of laboratory-based processes. These frameworks guide the user in the translation of lab operations into industrial process units, by providing simple scale-up formulae and leading the LCA practitioner in the analysis of similarities with existing industrial processes.

Another aspect that is generally overlooked in a laboratory synthesis but needs to be addressed considering its prospective industrial counterpart is the fate of waste streams (solid residues, gaseous and liquid effluents). The management methods adopted at lab scale might not be extended to the industrial scale. The LCA analyst might refer to the Best Available Technique reference documents, BREFs (European IPPC Bureau 2019), for an overview of the state-of-the-art on air and water pollution control devices in the industrial sector of interest. In BREFs, technologies for pollutant abatement are reviewed systematically, providing quantitative information on expected removal efficiency and associated consumption of reactants and energy. With a similar aim, Li et al. (2018) developed modular LCIs for some standard air pollution control units in the petrochemical sector. Differently from flue gases, effluents like waste solvents or wastewater streams might sometimes undergo treatment in external facilities (e.g. municipal wastewater treatment plants and municipal solid waste incinerators). To keep track of these streams and the inputs/outputs associated with their treatment, multi-input allocation models for LCI generation have been developed (see, e.g. Köhler et al. 2007 for wastewater treatment processes in the chemical sector and Seyler et al. 2004 for waste-solvent incineration).

1.3.2.3 Process Simulation

Once the process is defined, its modelling can be assisted by process simulation tools. Several commercial chemical process simulation (CPS) software packages are available on the market. Relevant examples include Aspen Plus, CHEMCAD, HYSYS, and Pro/II (Foo 2017).

All the considerations made above are still valid here, but the use of a process simulator offers a series of advantages over the simple process design calculations mentioned above. Besides providing material and energy flows for each equipment, process simulation allows for a quick screening of alternative plant configurations. Sensitivity analyses around operating conditions can be performed to assess how assumptions on the process variables influence the inventory data, thus exploring the uncertainties associated with the generated data.

Another advantage of computer-based tools is the relative ease in studying opportunities of process integration. This includes both energy and material integration within a process, as well as 'total site' integration, if the modelled process is to be realized, e.g. in an industrial park sharing infrastructure and utilities between processes. Generally speaking, integration can decrease markedly the entity of input and output flows in LCI, although trade-offs may also take place when, for instance, the recycling of by-products requires energy-intensive operations (Papadokonstantakis et al. 2016). A notable example of the use of CPS software to explore process integration in an LCA viewpoint is the recent work by Lari et al. (2018), where detailed process simulation allowed assessing the potential environmental advantages of integrating different glycerol upgrading routes within a single biorefinery.

The flexibility of CPS software is increased by the possibility to integrate external models within the framework of process simulation. For example, with the aim to determine the environmental footprint of the use of different solid sorbents for the abatement of gaseous acid pollutants, Dal Pozzo et al. (2017) nested a detailed reaction model inside a simulation of the acid gas treatment system in a CPS environment, thus coupling an accurate description of the gas–solid reaction with the modelling of unit operations and utilities provided by the CPS software. Likewise, the outputs of a process simulation can be interfaced with other, ancillary software tools. For example, the US EPA offers a suite of free tools for the estimation of uncontrolled (diffuse and fugitive) emissions from chemical processing equipment like, e.g. TANKS for the modelling of storage emissions (US EPA 2006a) or WATER9 for the modelling of fugitive emissions from wastewater treatment operations (US EPA 2006b), which can be complemented to process simulation.

1.3.3 Dedicated LCI Software

As the bottom-up approach relies on process design methods and it requires an extensive engineering knowledge that can be challenging, in addition to the above-described CPS software tools, dedicated software and/or tools have been developed in the past few years in order to support designers, engineers and LCA practitioners in view of establishing LCI data of chemical substances. A thorough web search, using search terms such as 'Life Cycle Inventory and tool' and 'Inventory and chemicals' has been carried out, but revealed a limited number of suitable tools (listed in Table 1.2).

CLiCC The Chemical Life Cycle Collaborative (CLiCC) tool is composed of three modules addressing different goals and needs of its user: (i) a screening-level assessment, (ii) a 'full' Life Cycle Assessment and (iii) Risk Assessment. With its first module, CLiCC enables a screening of life-cycle impacts for new chemicals, and thus can be used as early predictive tool, offering at the same time support for alternative comparisons. The second module (i.e. LCA) contains two sub-modules about life cycle inventory (LCI) estimates and life cycle impact assessment (LCIA) estimates. The first sub-module covers 'cradle-to-gate' inputs/outputs of chemical production via a customizable general chemical production model. The tool contains already more than 1.100 chemical-manufacturing process modules. Their input data—collected from publicly available industrial reports and general references such as Ullmann's Encyclopedia of Industrial Chemistry (Elvers 2011)—have been converted into LCI data. These LCI data have been established on the basis of stoichiometric equations in order to provide estimates of raw materials, efficiency, and product selectivity. The related energy requirements of each manufacturing process (for cooling or heating, as well as the energy gained/lost during

Table 1.2 Overview of software tools that support the establishment of LCI data of chemical substances

Tool	Goal and scope	Weblink
CliCC	Life-cycle impacts for chemical and materials at an early stage of the chemical product development process. The tool supports: (i) screening level assessment, (ii) Life Cycle Assessment and (iii) Risk Assessment	https://clicc.net
EATOS	Environmental assessment of organic syntheses to drive 'green' improvement of chemical synthetic sequence	http://www.metzger.chemie.uni-oldenburg.de/eatos/
Ecosolvent	Comparative environmental assessment of treatment technologies for specific, user-defined, waste-solvent mixtures	http://www.sust-chem.ethz.ch/tools/ecosolvent
Fine Chem	Estimation of the resource use and environmental impacts of petrochemical production based on molecular structure. The tool supports: (i) screening LCA, (ii) supply chain management and (iii) process design	https://www.ethz.ch/content/specialinterest/chab/chemicalnbiengneeing/setgroup/en/research/downloads/softwaretools/fine-chem.html
WW LCI 3.0	Life cycle inventories for urban wastewater discharges	https://lca-net.com/projects/show/wastewater-lci-initiative/

the different reaction steps) are established using thermodynamic theories (such as Gibbs free energy). The emissions were estimated using a chemical factor model. For new substances, the user can choose to use one of the existing data as 'proxy' chemical or to build a 'new' LCI, based on own input data. For the latter, reactants, co-products, by-products by mass, as well as the electricity and heating values are required. Artificial Neural Network (ANNs) is applied to characterize the relationship between chemical descriptors and life cycle impacts (CED, Acidification, GWP, Eco-indicator, Human health and ecosystem quality) (Song et al. 2017).

EATOS The Environmental Assessment Tool for Organic Synthesis (EATOS) is an easy-to-use tool that can be used to discuss and compare chemical reactions to their potential environmental impact to make them more environmentally friendly (Eissen and Lenoir 2017). The inventory data are established on the basis of the stoichiometry of the synthesis of the chemical under investigation. The tool provides a default list of synthesis mechanisms, however new synthesis routes can be added by the user. Only the input of the stoichiometric reaction equation and the amount of starting materials are required for the determination of the mass balancing. The tool comprises a weighting function that allows accounting the relevant environmental aspects for an estimation of the synthesis' danger potential. Quantitative material flows are weighted with their qualitative characteristics or weighting factor ('Q'). The tool comprises the claiming of resource consumption, risks, human toxicity, chronic toxicity, ecotoxicity, ozone creation, air pollution, accumulation, greenhouse effect, eutrophication and acidification as impact categories. For each category, a specific weighting factor (the values range from 0 to 10) is provided. Chemical data such as MAK value, hazard symbol, LD50, LC50, WGK value, POCP, BCF and R-phrases are used to determine the Q factors. The software provides a graphical evaluation that is useful for understanding which is the phase that has more environmental impacts, as well as for comparing different processes.

Ecosolvent This tool from ETH Zürich is a dedicated tool for the environmental assessment of waste-solvent treatment in the chemical industry. It is valuable to support a decision in chemical industry based on a retrospective assessment of the environmental impact of waste-solvent processes (Capello et al. 2007, 2008; Amelio et al. 2014). The tool comprises four different LCI models representing the most commonly used waste-solvent technologies—i.e. solvent recovery by distillation, thermal treatment in a cement kiln, waste-solvent incineration and wastewater treatment in case of aqueous distillation residues (Capello et al. 2005; Seyler et al. 2005). The integrated LCI model calculates waste-solvent specific inventory flow as a function of the waste-solvent composition and the treatment technology. The tool has a tiered structure; precise information on the treatment technologies, i.e. measured steam consumption of a distillation process, can be entered and the results show a relatively small uncertainty range. However, the tool may also be used if such precise information is missing and generic data are used to abridge the lack of information. In this case the results show larger uncertainty ranges. The model relies on the principle of a multi-input allocation model—for a detail explanation see Azapagic and Clift (1999)—allowing for '*a calculation of the environmental impacts of a specific product out of measurement data for a mixture of several products*' (Seyler et al. 2005). The statistical evaluation provides an empirical average value as well as the fitting of probability distributions for all the inventory parameters; these data can be used to perform a quantitative uncertainty analysis (i.e. Monte Carlo simulation). Inventory data for ancillaries, fuels, energy from industry data and the ecoinvent database are applied.

Fine Chem This tool from ETH Zürich is a dedicated tool to estimate the resource use and environmental impacts of petrochemical production (Wernet et al. 2008, 2009). The tool can be applied for the purposes of screening LCA, supply chain management and process design. The tool estimates the inventory data, of chemicals' production stage, based on the molecular structure of a chemical. The tool has been developed by using neural network model to estimate the key production parameters directly from the chemicals' molecular structure. The neural network model was established on mass and energy flow data on petrochemicals productions of 338 chemicals, based on industrial production data and on the ecoinvent database. The molecular structure-based model offers several advantages, such as, a low number of inputs data and a priori knowledge of the production process is not required. As an input ten chemical descriptors are required (i.e. number of nitrogen and halogen atoms, molecular weight, etc.). Results are cradle-to gate inventory data, as well as related Life Cycle Impact Assessment results (i.e. CED, GWP, Eco-indicator 99 score). Also, the tool provides an uncertainty analysis of the estimated values, allowing an assessment of their prediction accuracy.

WW LCI v3.0 A further tool (programmed in MS Excel) that is designed to calculate LCI of wastewater discharges down to the drain or directly into the aquatic environmental (Munoz et al. 2017; Kalbar et al. 2018; Munoz 2019). It provides a database on the wastewater treatment levels and sludge disposal practices of 81 different countries. Its database has been established on public data sources (i.e. Eurostat, OECD statistic and country-specific statistic). The Inventory data are calculated including infrastructure requirements, energy consumption, auxiliary materials for the treatment of wastewater, disposal of sludge and sewage. Four wastewater treatment levels are accounted: primary treatment (suspended solid settling), secondary treatment (aerobic biological treatment), tertiary treatment (nitrogen and phosphorus removal, sand filtration, disinfection by chlorination) and septic tank (on-site primary treatment). The outputs of wastewater level treatments are, i.e. fraction of chemical of effluent discharged in natural water body and the fraction of chemical discharged in the sludge. Sludge treatments by means of anaerobic digestion, composting, incineration and landfilling are modelled. The sludge treatments processes are assessed by mass balance equation (i.e. conversion of sludge into compost, combustion of biogas), and the data for the infrastructures are based on the ecoinvent data set (Munoz et al. 2017). In the model proposed by WW LCI, the fate of chemical (i.e. degradability) is included. Thus, the exchanges with the environment are assessed on the basis of the predicted behaviours (fate factor values) of the chemical in the wastewater treatment plant (WWTP) and in the environment. The fate factors of chemicals in WWTPs are assessed using SimpleTreat model (Franco et al. 2013) and USES-LCA (Van Zelm et al. 2009) is used to estimate the fate of the chemical in the environment.

Table 1.3 (on the next two pages) summarizes the key aspects of these various tools described in the preceding paragraphs above.

1.4 Top-Down Approach

Top-down approaches start, opposite to the above-described bottom-up approaches, from the most general level and proceed towards a much more (case-study) specific level. In an LCI analysis, a top-down approach uses general data (e.g. at global or national level) to derive the input and output flows definition of a specific system.

Table 1.3 Key aspects of the here examined tools, supporting the creation of LCI data from chemical synthesis processes

	[1] Area of application [2] Covered life stages	LCI based principle	Input parameters	Output parameters	LCIA output
Fine Chem	[1] Petrochemical, organic chemicals [2] Cradle-to-gate product synthesis	Molecular Structure-Based Models of Chemical Inventories using Neural Networks	Chemicals' descriptors: molecular weight, no. of functional groups, no. of oxygen atoms in keto and aldehyde groups, no. of nitrogen -halogen atoms, no. of aromatic-aliphatic ring, no. of ternary -quaternary carbon atoms, no. of heteroatoms in rings, number of unique substitutes on aromatic rings	Estimation of the key inventory parameters Cradle-to-gate inventory data	CED, GWP, Eco-indicator 99 scores, Electricity use and Heat use
EATOS	[1] Organic chemicals [2] Cradle-to-gate synthesis	Stoichiometric equations of synthesis processes	Mass and specific information which can be accessed by MSMD database	Quantity and yield of product, coupled products, by-products, auxiliary materials	Environmental Index (EI)
CliCC	[1] Organic chemicals [2] Cradle-to-gate synthesis and separation processes	Stoichiometric equations and thermodynamic theories	Raw material and energy inputs necessary for chemical synthesis and separation. Default database library (LCI for platform chemicals, process parameters for unit operations, use and end-of-life parameters). User input: direct emission, user-defined processes and scenario)	Estimation of total input and emission for 1 kg of chemical	Artificial Neural Networks (ANNs) is used to identify the relationship between chemical descriptors and life cycle impacts. Impact categories: CED, Acidification, GWP, Eco-indicator point, Human health, Ecosystem quality

(continued)

Table 1.3 (continued)

	[1] Area of application [2] Covered life stages	LCI based principle	Input parameters	Output parameters	LCIA output
Ecosolvent	[1] Organic chemicals [2] Waste-solvent treatment processes: thermal treatment in hazardous waste-solvent incinerators and in cement industry, solvent recover (solvent distillation), wastewater treatment. Transport of chemical, construction and maintenance of building/infrastructure/technical equipment are neglected	Multi-input allocation method	Thermal treatment in cement kilns: amount, net calorific value and elementary composition of waste solvent, energy, ancillary materials. Thermal treatment: feed water, energy, ancillary materials Solvent distillation: steam, electricity, cooling water, inert gas, ancillary products, recovered solvent, residues, wastewater, outlet air. Wastewater treatment (gate-to-gate): energy, auxiliaries, chemical	Emissions of pollutants into the environmental media, generation of residues and of co-products	CED, GWP, Eco-indicator point, Ecological scarcity
WW LCI 3.0	[1] Organic and inorganic chemicals [2] WWTP primary-secondary-tertiary-treatment, sludge disposal, chemicals' degradation in the environment	Mass balance; Generating inventories for chemicals with inclusion of environmental degradation	Chemical-specific variables (physical-chemical properties, elementary composition, fate factors in WWTP and in the environment) and scenario variables (% of population connected to WWTPs, nutrients removal, sludge disposal scenario; inventory of infrastructure, energy consumption)	Emission to nature from WWTP, from environmental degradation, from sludge composting, from sludge landfilling, from sludge incineration, from sludge application to soil Emission to technosphere from WWTP, from sludge landfilling, from sludge incineration, from sludge application to soil	A comprehensive inventory linked to ecoinvent v3

1.4.1 Economic Input–Output Analysis

Economic input–output (EIO) analysis is a macroeconomic technique where the complex interdependencies across different sectors/branches of an economy are represented by a set of linear equations. EIO analysis was theorized and developed by Leontief in the 1930s (Leontief 1936). The most basic form of EIO analysis involves so-called input–output tables. Such tables quantify the supply chain for all sectors of an economy displaying information using a series of rows and columns of data. Industry sectors are listed in the heads of each row and each column. The data in each row represents the amount of output sent from an industrial sector to the different column sectors for productive use. The data in each column corresponds to the amount of inputs used in that industrial sector.

In the 1970s, along with increasing concerns on the environment, Leontief and co-authors proposed to include environmental externalities in input–output models. In such a way, pollutant emissions and natural resources consumption are treated as sectors of the input–output model (Leontief and Ford 1970). In 1995, Lave and co-workers proposed the economic input–output life-cycle analysis (EIO-LCA) to address the problem of subjective boundary definition (Lave et al. 1995). The EIO-LCA models incorporate economic matrices of sector-based environmental and resource-use coefficients. Those models allow the LCA researchers and practitioners to evaluate the environmental impacts associated with a product, as represented by one or more economic sectors. Since the initial implementation in the US (Hendrickson et al. 1998), EIO-LCA models have been developed for many countries around the world and have been widely used for analysing a wide range of products, including fuels, chemicals, and plastics, as described in the following.

1.4.1.1 Mathematical Structure of Economic Input-Output Life-Cycle Assessment

The essential tool for the analysis is a matrix indicating sector-to-sector flows of purchases, which is denoted by **A** and called the direct requirements coefficient matrix. An element a_{ij} of matrix A represents the monetary value of the input required from sector i to produce one unit of monetary output of sector j ($i = 1 \ldots n$, and $j = 1 \ldots n$) (e.g. value of electricity required to produce one monetary unit of steel). **I** is the identity matrix of dimensions n by n (to account for the output of each sector production stage itself). Finally, **Y** represents the vector of the required output of each sector (e.g. 100 dollars of steel) and **X** represents the vector of total inputs of each sector (e.g. dollars of electricity required to produce 100 dollar of steel). Then **X** can be obtained by multiplying the matrix (**I** + **A**) by the vector of required outputs:

$$\mathrm{X} = (\mathrm{I} + \mathrm{A})\mathrm{Y} \tag{1.1}$$

In (1.1) only direct (first-level supplier) inputs are taken into considerations. It is possible to consider also the second-level supplier requirements considering the matrix **AA**, the third-level supplier requirements by the **AAA** matrix, etc. Therefore, the X vector including all supplier outputs can be obtained as:

$$X = (I + A + AA + AAA + \ldots)Y \quad (1.2)$$

The expression (**I** + **A** + **AA** + **AAA** + …) can be shown to be equivalent to $(\mathbf{I} - \mathbf{A})^{-1}$; then (1.2) can be written as:

$$X = (I - A)^{-1}Y \quad (1.3)$$

The $(\mathbf{I} - \mathbf{A})^{-1}$ matrix is commonly named Leontief Inverse matrix.

This EIO analysis can be adapted to environmental purposes. An environmental impact would characterize the discharges into air, water, underground, and on land of an output from each industry. Suppose B is a k by n matrix of environmental burden coefficients, where b_{kj} is environmental burden k (e.g. carbon monoxide emissions) per dollar output of sector j; and **M** is the vector of total environmental burdens. Then, the vector of total environmental burdens is calculated by:

$$M = B(I - A)^{-1}Y \quad (1.4)$$

1.4.1.2 EIO-LCA Applied to Chemicals

Lave et al. (1995) proposed the first EIO-LCA approach with the aim of capturing economic interdependencies for examining the economy-wide environmental implications of a product. The authors compared plastic cups to paper cups and they applied the US economic input–output tables that were compiled for 519 sectors. The two products have been approximated by the corresponding commodity sectors. In detail, they modelled the paper cups by using the data for the industry sector 'Paperboard Containers and Boxes' and the plastic cups by using the industry sector 'Plastic Materials and Resins'. The electricity consumption and the toxic chemicals release have been used as environmental burdens. The toxic chemicals release was characterized by destination media (air, water, underground, land) and by substances (320 toxics were included). Results highlighted the relevance of indirect suppliers that are generally neglected by 'conventional' (bottom-up) LCI.

Joshi (1999) proposed an analytical model consisting of the 498 sectors of US economic input–output tables augmented with various sector-level environmental burden vectors, including energy use, non-renewable ores use, conventional pollutant emissions, toxic releases, hazardous solid waste generation and fertilizer use (as an indicator of the eutrophication potential). Impacts from individual pollutant emissions were aggregated using appropriate characterization factors in global warming, acidification, ozone depletion, toxicity and energy use. In addition to the EIO-LCA approach, Joshi (1999) proposed five alternative models that (i) introduce a new hypothetical sector entering into the economy (Model II); (ii) use selective disaggregation of aggregate input–output data (Models III and IV); (iii) include use and end-of-life phases (Models V and VI). These alternative models were, subsequently, classified as 'hybrid approaches'. The author presented a case study comparing steel and plastic automobile fuel tank systems and he also compared the results of EIO-LCA to a conventional process-based LCA. The author highlighted that despite using very different methodologies there is a high degree of correspondence between the comparable results from these two approaches. Moreover, the absolute amount of impacts reported in EIO-LCA is generally lower than those reported by the process-based approach (Joshi 1999).

In 1999, the EIO-LCA method was transformed into a user-friendly online tool by researchers at the Green Design Institute of Carnegie Mellon University. The website is free for non-commercial use. The EIO-LCA models available on the site apply the EIO-LCA method to various national and state economies. Each model is comprised of national economic input–output models and publicly available resource use and emissions data. Since 1999, the online tool has been accessed over 1 million times by researchers, LCA practitioners, business users, students and others (CMU Green Design Institute 2019).

While the method found applications in many countries (e.g. Australia, Germany, Denmark, Japan), in the USA the efforts in the EIO-LCA development continued and permitted great progresses. In 2005, a new environmental database for the US, named 'Comprehensive Environmental Data Archive (CEDA) 3.0' was launched (Suh 2005). CEDA 3.0 covered a total of 1,344 environmental flows (both resource and emissions). The toxic pollutants part of the database comprises about one thousand toxic chemicals. Data derived from Toxics Releases Inventory, National Toxics Inventory and National Center for Food and Agricultural Policy databases.

The next step was made by Hawkins and co-authors (Hawkins et al. 2007). The authors combined Materials Flow Analysis (MFA) and EIO model to create the Mixed-Unit Input-Output (MUIO) model. In their approach the sector output, which is generally considered in monetary terms, can be also expressed in physical units. Therefore, the MUIO model provides the total physical and monetary output required to meet an additional final demand by sector and stage, in the supply chain. In their study, they tested the model by using physical and monetary units to describe the output of sectors that produce heavy metals and products containing these metals. Material flow data were obtained from the US Geological Survey.

Generally, EIO-LCA allows evaluating the environmental impacts of chemical emissions just on the basis of the overall mass of chemicals released into the environment, without taking into consideration the potential concentration of those chemicals in the different environmental media. This is performed through the data supplied by individual plants of specified industries about their toxic releases into air, water, land and underground. Wright et al. (2008) integrated EIO-LCA with a multimedia fate and transport model (CHEMGL) and a human risk assessment tool for a screen-level-analysis to examine the relative risk posed during each life cycle stage of a chemical. The CHEMGL model was applied to predict the chemical concentration in environmental compartments, for each life cycle stage of a chemical (the production, manufacturing and consumption stages). The predicted concentration is then input into a risk assessment tool. The latter assesses the human and fish risk potential, per unit of chemical released, by combining toxicity data (i.e. reference dose) and exposure via inhalation and ingestion. Finally, the relative risk, in units of toluene equivalent, is determined by multiplying the risk potential per unit release by the environmental release into the air boundary layer, surface water, surface soil and groundwater based on EIO-LCA model. This integrated life cycle methodology allows chemical designers to evaluate each stage and assess areas where the risk can be minimized using alternative chemicals or process operations.

Recently, Meng and Sager (2017) provided a global view of China's petrochemical industry's energy consumption and CO_2 emissions by using an EIO-LCA model from both production and demand perspectives. The method permits the authors to calculate not only the direct energy consumption and energy-related CO_2 emissions but also the indirect amounts that can be identified throughout the supply chain. The results of the study indicate that the indirect energy consumption and CO_2 emissions embodied and conveyed throughout the supply chain are highly significant, though often overlooked. The Chinese petrochemical industry accounts for 23% of the total energy consumption and 32% of CO_2 emissions. These indirect loads cannot be neglected in determining the role and responsibilities of sectors in energy conservation and emissions reduction.

The EIO-LCA addresses the conventional LCA limitation of excluding a large set of activities from the analysis with narrow study boundaries since the entire national economy is included in the boundary of this analysis. Moreover, it offers the advantage that analyses can be performed in minutes, not months. Also, all data used by the model are publicly available government data and therefore valid and accepted. The main limitation of EIO-LCA analysis is the problem of data aggregation, resulting in a limited granularity: the product of interest, indeed, is approximated by its commodity sector in the national input–output tables with respect to input requirements; in this way the inventory data suffers from scarce accuracy.

1.4.2 Data Mining, Web Mining and Big Data Applied to Chemicals

According to Hand et al. (2001), 'Data mining is the analysis of (often large) observational data sets to find unsuspected relationships and to summarize the data in novel ways that are both understandable and useful to the data owner'.

Data mining could be another suitable approach to compile chemical inventories (Cashman et al. 2016). Relevant examples in chemical fields are the exploitation of the national inventories to develop EIO-LCA models. For example, Suh (2005) integrated US EPA Toxics Release Inventory (TRI) and National Emissions Inventory (NEI) to develop sector-average emission data for all industries covered in the EIO tables, including petrochemical production. Recently, Sengupta et al. (2015) highlighted that the potential for underreporting in national emission inventories can result in significant potential for underestimation of emission factors when calculated on a sector-average basis. In this regard, the authors proposed an approach that can be used to improve the estimation of sector-average emission factors for use in EIO-LCA studies and, starting from national inventories, used it for a case study estimating environmental impacts of ethanol and gasoline.

Data mining is an important approach also for gathering characterization data. For example, using the physicochemical and toxicity data available in databases from the European Chemical Agency, the European Food Safety Authority and the University of Hertfordshire, EC-JRC has calculated ecotoxicity and toxicity characterization factors for over 6,000 chemical substances (Saouter et al. 2018).

Web mining is the use of data mining to discover and extract information from web pages. In the last few years, many national statistical institutes provide open data concerning several specific sectors (e.g. environment, economy, society, etc.) that could be extremely useful for implementing LCA models. Often such data have different formats and are incomplete and it is therefore difficult to share and compare them. In order to overcome these problems, the Semantic Web proposes standards to promote common data formats and exchange protocols on the web (Berners-Lee et al. 2001). The US EPA is implementing semantic management of linked open data (LOD) for environmental databases such as Toxics Release Inventory (TRI) and the Chemical Data Reporting tool (CDR). This shift can support a transition to semantic data mining for improved inventory modelling. In this sense, very recently, some scientific papers analysed the use of semantic information resources for improving LCA (Cashman et al. 2016; Ingwersen et al. 2015; Kuczenski et al. 2016; Zhang et al. 2015). In the field of chemical products and processes, the work of Cashman et al. (2016) is of special interest. These authors have proposed a method for standardizing and automating the discovery and use of publicly available data at the US EPA for chemical-manufacturing LCI. The method is applicable to chemicals traced in CDR. In 2018, Mittal and co-authors presented two coupled ontologies, i.e. semantic data models, with the ultimate goal of developing an automated life cycle inventory modelling (Mittal et al. 2018). The ontologies, called Lineage and Process, were developed in the Web Ontology Language. The described ontological modelling provides a means to identify and eventually predict the synthesis route of a chemical, while connecting this knowledge with relevant process information.

The definition of Big Data generally refers to large and unwieldy groups of data that regular database management tools have difficulty in capturing, storing, sharing and managing (Sharma and Gulia 2014). Recently, the use of Big Data to collect a great amount of data for implementing LCI has prompted much interest in the LCA community. Cooper et al. (2013) reported several examples of Big Data collected by national agencies and institutions and used in LCI and LCIA. Song et al. (2017) analysed the use of Big Data in environmental performance evaluation with a focus on thermal power plants. The authors highlighted that combined use of Big Data and LCA should be significantly improved before being successfully applied to assess environmental performances.

1.5 Hybrid Approach Applied to Chemicals

Hybrid techniques attempt to combine the benefits of both process-based and EIO analyses, while minimizing their limitations. Process-based analysis is accurate and detailed but is generally time-intensive and can be subject to considerable variability. EIO analysis is relatively fast and representative at the national level; moreover, the system boundaries are wide-ranging; however, it can be less accurate than process-based LCI (Treloar 1997). As stipulated above, Suh and Huppes (2005) distinguished three categories of hybrid analyses: (i) tiered hybrid or process-based hybrid; (ii) IO-based hybrid and (iii) integrated hybrid. Tiered hybrid analysis utilizes process-based inventory data for use and disposal phases, as well as for several important upstream processes, and then the remaining input data are imported from an EIO-based LCI (Suh and Huppes 2005). An example of its first application is the Model II proposed by Joshi (1999). Then, in order to improve the tiered hybrid analysis, Suh and Huppes (2002) introduced the Missing Inventory Estimation Tool (MIET) that combines the strengths of process-based LCA and EIO-LCA. The general strategy of MIET is to minimize the use of EIO tables for major processes by restricting its application only to the flows located at the margin of the system boundaries. IO-based hybrid analysis is based on the extraction of a particular path from EIO table and substituting them with process-based data (Islam et al. 2016). The disaggregation procedure is the most essential part of IO-based hybrid approach. Models III and IV proposed by Joshi (1999) concern disaggregation procedures in hybrid approaches. An integrated hybrid approach implies that the process-based data is fully incorporated into the IO model, represented in a technology matrix by physical units per unit operation time of each process while the input–output system is represented by monetary units (Suh et al. 2004).

Hybrid approach has been applied successfully also to the chemical sector. Biofuel sector shows several applications of hybrid LCI analysis. Life cycle water consumption of corn-based ethanol, soybean biodiesel, cellulosic ethanol from switch grass and microbial biodiesel was assessed by Harto et al. (2010). The tiered hybrid model was applied due to the wide range of technologies being covered. Tiered hybrid analysis was used also by Strogen and Horvath (2013) in order to assess life-cycle greenhouse gas (GHG) emissions of petroleum and biofuels. Hybrid LCI analysis detailed physical data were used as inputs to model energy production and consumption processes, and economic data were used to characterize manufacturing, construction and maintenance activities. In Watanabe et al. (2016) a hybrid approach was applied to the first- and second-generation ethanol production in Brazil. The method is similar to IO-based hybrid analysis: data on direct and downstream requirements are collected according to the process analysis, whereas the remaining upstream requirements are covered by EIO tables. Data for each biorefinery from process-based LCI were inserted as a group of new sectors and commodities into the IO model by following the approach of Joshi (1999). A very promising approach is the combined use of multi-objective optimization and integrated hybrid LCA (Yue et al. 2016). This quantifies both direct and indirect environmental impacts and incorporates them into the decision-making process in addition to the conventional economic criteria. That approach has been demonstrated in the bioethanol supply chain. Liu et al. (2018) have applied the tiered hybrid model to the environmental assessment of biofuels from corn-stover: the impacts caused by the direct chemical emissions are evaluated by process-based approach, whereas the environmental impacts caused by the indirect chemical emissions are analysed by the EIO-LCA model. The indirect emissions are found to be significant.

Two further examples of hybrid approach applied to chemical products are the studies of Tatari et al. (2012) and Rodríguez-Alloza et al. (2015) on warm-mix asphalts. The former study developed a tiered hybrid LCA model and focused on the thermodynamic aspects, the latter one performed an IO-based hybrid analysis.

A final example is the work of Alvarez-Gaitan et al. (2013), which has developed a tiered hybrid LCA of water treatment chemicals. The authors compared these results with process and input–output models for caustic soda, sodium hypochlorite, ferric chloride, aluminium sulphate, fluorosilicic acid, calcium oxide and chlorine gas. In many cases, very close results have been obtained from process-based- and hybrid-LCA. Moreover, the research shows that where there are important price fluctuations in the raw materials, hybrid modelling provided a more robust output.

Discussion and Conclusions

Existing life cycle databases cover only a portion of the vast and increasing variety of chemicals available on the market. LCA practitioners willing to investigate the environmental footprint of novel products and novel production routes have the need to obtain first-hand representative data on the related material and energy flows. This chapter offered a concise overview of the possible approaches to generating inventory data.

In bottom-up approaches, the LCI is developed from process considerations, ranging from the mere knowledge of reaction stoichiometry to a full-fledged process simulation, passing through process design calculations of gradually increasing complexity. A clear advantage of this approach is the control on the modelled system. For the LCA analyst that develops a process-based LCI, the gate-to-gate production process of a target chemical is not anymore a non-detailed 'black box', with just known inputs and outputs, as in the traditional LCA practice. As such, LCIs generated from process-based considerations are potentially more detailed than LCIs found in commercial databases and offer to their creators the flexibility to assess changes in the process conditions and an increased awareness of the uncertainties in the inventory data.

Clearly enough, increasing the level of detail in LCI generation comes at the price of increasing time effort and increasing requirements of domain knowledge in industrial chemistry and chemical engineering. In particular, process modelling and simulation is a powerful approach, but solid expertise is required to generate meaningful, representative LCI results. The trade-off between accuracy in LCI modelling and time/expertise constraints has to be clear to the LCA practitioner. Not every LCA study requires the highest level of detail in inventory compilation and, depending on the specific goal and scope, the choice of simpler process-based methods can be more than adequate.

In this respect, dedicated LCI software tools have been developed to streamline the compilation of inventories for specific systems. The development of these tools is a rather new area of LCA-related research. In fact, only 5 dedicated LCI software could have been identified for the area of synthesis of chemical products. These tools are not in all cases covering all the life cycle stages of the actual chemical synthesis—i.e. some of them focus only on a single stage (e.g. discharges) of the entire life cycle of a chemical. One part of these tools also include an LCIA module and therefore allows a full LCA computation of (the synthesis of) chemical products. These dedicated tools are particularly useful in screening applications. When a vast number of alternative chemicals has to be compared to identify the 'greener' choice (e.g. for a solvent), it would be unfeasible to elaborate process flow diagrams for the production of each compound, whereas the use of a dedicated tool can give a swift, approximate result. On the other hand, care should be given in using such tools properly, for example avoiding extrapolation to product categories/production routes not covered by the software (e.g. using for the production of a chemical via a bio-based route a tool trained only for conventional, fossil-based processes).

At the other end of the spectrum compared to bottom-up methods, top-down approaches use general data to compile LCI for a specific system. EIO-LCA takes a top-down approach and treats the whole economy as the boundary of analysis, thus overcoming the conventional LCA limitation of excluding a large set of activities from the analysed system because of a narrow boundary definition. EIO-LCA offers the advantage that analyses can be performed very quickly. Also, all data used by the model are publicly available government data; therefore, the issue of preserving the confidentiality of industrially sensitive information does not exist. The main limitation of EIO-LCA analysis is the problem of aggregation. Therefore, the product of interest has to be approximated by its commodity sector in the national input–output tables with respect to input requirements and environmental coefficients. Moreover, EIO-LCA accounts the upstream environmental burdens associated with raw materials extraction and manufacturing phases, but not those associated with use and end-of-life options. Lastly, EIO-LCA allocates environmental burdens based on market value.

Web mining will continue and even increase over the coming decades and it shows enormous potential for applications in LCA. Currently, there is the need of standardization of format and the Semantic Web offers a common framework for standardizing data. Big Data appears as a vital source of data and the LCA community has already started to use it. However, this open and collaborative model for data production creates new challenges in data integration and harmonization. Therefore, there is also a need to verify applicability, reliability and stability of an application of Big Data in the assessment of environmental performances. All of this is fundamental in order to provide environmental performance assessments by using Big Data in order to provide a scientific basis for supporting environmental management policies. However, to date, a significant improvement in the integration of Big Data and LCA is needed for their successful application in environmental performance evaluations.

A hybrid approach combines the strengths of top-down input–output analysis and bottom-up process analysis, allowing for specificity, accuracy and system completeness while eliminating boundary truncation errors. That approach allows direct and indirect burdens to be included and some authors demonstrate that indirect burdens are often not negligible. Free online input–output tools, like EIO-LCA (Carnegie Mellon University Green Design Institute 2012), would be helpful in the future development of hybrid models.

With such a variety of methods available for LCI compilation, the LCA analyst has to make a choice based on the specific study to be developed. Meyer et al. (2019) recently introduced the concept of 'purpose-driven reconciliation' to propose a logical framework for approach selection. Depending on the goal and scope of the LCA study to perform (e.g. whether it has a regulatory or design purpose, whether it assesses a novel or existing substance, whether it is a screening of alternative chemicals or a detailed comparison of process schemes, etc.), the most important constraint can be complying with a data quality criterion, or meeting the timeline to decision, or addressing specific data gaps. By analysing the trade-offs of each available approach, as has been briefly done in this chapter, the LCA practitioner should identify which of these generally meets the constraints and opt for it, while leaving open the possibility of using alternative methods to retrieve specific missing information or even to revisit the decision-making approach, if needed.

References

Alvarez-Gaitan PJ, Peters MG, Rowley VH et al (2013) A hybrid life cycle assessment of water treatment chemicals: an Australian experience. Int J Life Cycle Assess 18:1291–1301

Amelio A, Genduso G, Vreysen S et al (2014) Guidelines based on life cycle assessment for solvent selection during the process design and evaluation of treatment alternatives. Green Chem 16:3045

Arvidsson R, Molander S (2016) Prospective life cycle assessment of epitaxial graphene production at different manufacturing scales and maturity. J Ind Ecol 21:1153–1164

AusLCI (2019) Australian Life Cycle Inventory Database Initiative. http://www.auslci.com.au. Accessed 30 May 2019

Azapagic A, Clift R (1999) Allocation of environmental burdens in multiple-function systems. J Cleaner Prod 7:101–119

Berners-Lee T, Hendler J, Lassila O (2001) The semantic web. Sci Am 284:34–43

Boustead I (2005) Eco-profiles of the European plastics industry. Association of Plastics Manufacturers in Europe (Plastics Europe), Brussels

Caduff M, Huijbregts MAJ, Koehler A et al (2014) Scaling relationships in life cycle assessment. J Ind Ecol 18:393–406

Capello C, Hellweg S, Badertscher B et al (2007) Environmental assessment of waste-solvent treatment options. J Ind Ecol 4:26–38

Capello C, Hellweg S, Badertshcer B et al (2005) Life cycle inventory of waste solvent distillation: statistical analysis of empirical data. Environ Sci Technol 39:5885–5892

Capello C, Hellweg S, Hungerbühler K (2008) Environmental assessment of waste-solvent treatments options. J Ind Ecol 12:111–127

Carnegie Mellon University (CMU) Green Design Institute (2019) Economic input-output life cycle assessment (EIO-LCA) model. CMU, Pittsburgh. http://www.eiolca.net. Accessed 28 Feb 2019

Cashman SA, Meyer DE, Edelen NA et al (2016) Mining available data from the United States environmental protection agency to support rapid life cycle inventory modeling of chemical manufacturing. Environ Sci Technol 50:9013–9025

Cefic (2018). FACTS & FIGURES of the European chemical industry. http://www.cefic.org/Documents/RESOURCES/Reports-and-Brochure/Cefic_FactsAnd_Figures_2018_Industrial_BROCHURE_TRADE.pdf. Accessed 28 Feb 2019

Chase WM (1998) NIST-JANAF Thermochemical Tables. American Institute of Physics for the National Institute of Standards and Technology, Washington, DC

Cooper J, Noon M, Jones C et al (2013) Big data in life cycle assessment. J Ind Ecol 17:796–799

CPM (2019) SPINE@CPM— the operational prototype for industrial LCA databases. Center for Environmental Assessment of Product and Material Systems, Chalmers University, Göteborg, Sweden. http://cpmdatabase.cpm.chalmers.se/AboutDatabase_1.htm. Accessed 30 May 2019

CRMD (2019) Canadian raw materials database. University of Waterloo. http://uwaterloo.ca/canadian-raw-materials-database/. Accessed 30 May 2019

Curran MA (2012) Sourcing Life Cycle Inventory Data. In: Curran MA (ed) Life cycle assessment handbook: a guide for environmentally sustainable products. Wiley, Hoboken, NJ p, pp 105–141

Dal Pozzo A, Guglielmi D, Antonioni G et al (2017) Sustainability analysis of dry treatment technologies for acid gas removal in waste-to-energy plants. J Cleaner Prod 162:1061–1074

de Camargo AM, Forin S, Macedo K et al (2018) The implementation of organizational LCA to internally manage the environmental impacts of a broad product portfolio: an example for a cosmetics, fragrances, and toiletry provider. Int J Life Cycle Assess 24(1):104–116

De Rosa-Giglio P, Fontanella A, Gonzalez-Quijano G, Ioannidis I, Nucci B, Brugnoli F (2018) Product environmental footprint category rules - Leather. https://ec.europa.eu/environment/eussd/smgp/pdf/PEFCR_leather.pdf. Accessed 19 Nov 2019

ECHA (2018) 21,551 chemicals on EU market now registered. https://echa.europa.eu/-/21-551-chemicals-on-eu-market-now-registered. Accessed 28 Feb 2019

Eissen M, Lenoir D (2017) Mass efficiency of alkene syntheses with tri-and tetrasubstituted double bonds. ACS Sustain Chem Eng 5:10459–10473

Elvers (ed) (2011) Ullmann's encyclopedia of industrial chemistry, 7th edn. Wiley-VCH, Weinheim

European IPPC Bureau (2019) Reference documents under the IPPC directive and the IED. http://eippcb.jrc.ec.europa.eu/reference. Accessed 28 Feb 2019

Foo D (2017) Chemical engineering process simulation, 1st edn. Elsevier, Amsterdam

Franco A, Stuikis J, Gouin T et al (2013) Evolution of the sewage treatment plant model SimpleTreat: use of realistic biodegradability tests in probabilistic model simulations. Integr Environ Assess Manage 9:569–579

Gavankar S, Suh S, Keller AA (2015) The role of scale and technology maturity in life cycle assessment of emerging technologies: a case study on carbon nanotubes. J Ind Ecol 19:51–60

Green D, Perry R (2007) Perry's chemical engineer's handbook, 8th edn. McGraw-Hill, New York, NY

Khoo HH, Isoni V, Sharratt PN (2018) LCI data selection criteria for a multidisciplinary research team: LCA applied to solvents and chemicals. Sustain Prod Consum 16:68–87

Hall S (2017) Rules of thumb for chemical engineers, 6th edn. Elsevier, Amsterdam

Hand D, Mannila H, Smyth P (2001) Principles of data mining. MIT Press, Cambridge, MA

Harto C, Meyers R, Williams E (2010) Life cycle water use of low-carbon transport fuels. Energ Policy 38:4933–4944

Hawkins T, Hendrickson C, Higgins C et al (2007) A mixed-unit input-output model for environmental life-cycle assessment and material flow analysis. Environ Sci Technol 41:1024–1031

Hendrickson C, Horvath A, Joshi S et al (1998) Economic input-output models for lifecycle assessment. Environ Sci Technol 13:184A–191A

Hischier R, Hellweg S, Capello C et al (2005) Establishing life cycle inventories of chemicals based on differing data availability. Int J Life Cycle Assess 10:59–67

Hischier R, Kwon NH, Brog J-P et al (2018) Early-stage sustainability evaluation of nanoscale cathode materials for Lithium ion batteries. Chemsuschem 11:2068–2076

Holmes JL, Aubry C (2012) Group additivity values for estimating the enthalpy of formation of organic compounds: an update and reappraisal. 2. C, H, N, O, S, and Halogens. J Phys Chem A 116:7196–7209

Huang H-J, Ramaswamy S, Tschirner UW et al (2008) A review of separation technologies in current and future biorefineries. Sep Purif Technol 62:1–21

Ingwersen WW, Hawkins TR, Transue TR et al (2015) A new data architecture for advancing life cycle assessment. Int J Life Cycle Assess 20:520–526

IDEA (2019) Inventory database for environmental analysis. National Institute of Advanced Industrial Science and Technology, Japan. http://www.idea-lca.jp. Accessed 30 May 2019

ISO (2006a) Environmental management—life cycle assessment—principles and framework. Geneva, International Standardization Organization (ISO), European Standard EN ISO 14040

ISO (2006b) Environmental management—life cycle assessment—requirements and guidelines. Geneva, International Standardisation Organisation (ISO), European Standard EN ISO 14044

Islam S, Ponnambalam SG, Loong Lam H (2016) Review on life cycle inventory: methods, examples and applications. J Cleaner Prod 136:266–278

Jiménez-González C, Kim S, Overcash MR (2000) Methodology for developing gate-to-gate life cycle inventory information. Int J Life Cycle Assess 5:153–159

Jiménez-González C, Overcash MR (2014) The evolution of life cycle assessment in pharmaceutical and chemical applications—a perspective. Green Chem 16:3392–3400

Joshi S (1999) Product environmental life-cycle assessment using input output techniques. J Ind Ecol 3:95–120

JRC (2019) Life cycle data network. Joint Research Centre, European Union. https://eplca.jrc.ec.europa.eu/LCDN/. Accessed 30 May 2019

Kalbar PP, Munoz I, Birkved M (2018) WW LCIv2: a second-generation life cycle inventory model for chemicals discharged to wastewater system. Sci Total Environ 622:1649–1657

Köhler A, Hellweg S, Recan E et al (2007) Input-dependent life-cycle inventory model of industrial wastewater-treatment processes in the chemical sector. Environ Sci Technol 41:5515–5522

Kroschwitz JI, Seidel A (eds) (2004) Kirk-Othmer encyclopedia of chemical technology, 5th edn. Wiley-Interscience, Weinheim

Kuczenski B, Davis CB, Rivela B et al (2016) Semantic catalogs for life cycle assessment data. J Cleaner Prod 137:1109–1117

Linstrom PJ, Mallard WG (eds) (2019) NIST Chemistry WebBook, NIST Standard Reference Database Number 69. National Institute of Standards and Technology, Gaithersburg, MD. https://doi.org/10.18434/T4D303. Accessed 28 Feb 2019

Lari GM, Pastore G, Haus M et al (2018) Environmental and economical perspectives of a glycerol biorefinery. Energy Environ Sci 11:1012–1029

Laurent A, Bakas I, Clavreul J et al (2014) Review of LCA studies of solid waste management systems—Part I: Lessons learned and perspectives. Waste Manage 34:573–588

Lave LB, Cobas-Flores E, Hendrickson CT et al (1995) Using input-output analysis to estimate economy-wide discharges. Environ Sci Technol 29:420A–426A

Leontief WW (1936) Quantitative input and output relations in the economic systems of the United States. Rev Econ Stat 18:105–125

Leontief WW, Ford D (1970) Environmental repercussions and the economic structure: an input–output approach. Rev Econ Stat 52:262–271
Li S, Feliachi Y, Agbleze S, Ruiz-Mercado GJ et al (2018) A process systems framework for rapid generation of life cycle inventories for pollution control and sustainability evaluation. Clean Technol Environ Policy 20:1543–1561
Liu C, Huang Y, Wang X et al (2018) Total environmental impacts of biofuels from corn stover using a hybrid life cycle assessment model combining process life cycle assessment and economic input-output life cycle assessment. Integr Environ Assess Manage 14:139–149
Meng L, Sager J (2017) Energy consumption and energy-related CO_2 emissions from China's petrochemical industry based on an environmental input-output life cycle assessment. Energies 10:1–12
Meyer DE, Mittal VK, Ingwersen WW et al (2019) Purpose-driven reconciliation of approaches to estimate chemical releases. ACS Sustain Chem Eng 7:1260–1270
Mittal V, Bailin S, Gonzalez M et al (2018) Toward automated inventory modeling in life cycle assessment: the utility of semantic data modeling to predict real-world chemical production. ACS Sustain Chem Eng 6:1961–1976
Moriguchi Y, Kondo Y, Shimizu H (1993) Analyzing the life cycle impact of cars: the case of CO_2. Ind Environ 16:42–45
Munoz I, Otte N, Van Hoof G et al (2017) A model and tool to calculate life cycle inventories of chemicals discharged down the drain. Int J Life Cycle Assess 22:986–1004
Munoz I (2019) Wastewater life cycle inventory initiative. WW LCI 3.0: changes and improvements to WW LCI v2. 2.-0 LCA Consultants, Aalborg
NREL (2019) U.S. Life cycle inventory database. National Renewable Energy Laboratory, Lakewood, CO, USA. https://www.nrel.gov/lci/. Accessed 30 May 2019
Papadokonstantakis S, Karka P, Kikuchi Y et al (2016) Challenges for model-based life cycle inventories and impact assessment in early to basic process design stages. In: Ruiz-Mercado G, Cabezas H (eds) Sustainability in the design, synthesis and analysis of chemical engineering processes. Butterworth-Heinemann, Oxford, p 295
Parvatker AG, Eckelman MJ (2019) Comparative evaluation of chemical life cycle inventory generation methods and implications for life cycle assessment results. ACS Sustain Chem Eng 7:350–367
Piccinno F, Hischier R, Seeger S et al (2016) From laboratory to industrial scale: a scale-up framework for chemical processes in life cycle assessment studies. J Cleaner Prod 135:1085–1097
Raymond MJ, Slater CS, Savelski MJ (2010) LCA approach to the analysis of solvent waste issues in the pharmaceutical industry. Green Chem 12:1826–1834
Righi S, Baioli F, Dal Pozzo A et al (2018) Integrating life cycle inventory and process design techniques for the early estimate of energy and material consumption data. Energies 11:970
Rodriguez-Alloza AM, Malik A, Lenzen M et al (2015) Hybrid input-output life cycle assessment of warm mix asphalt mixtures. J Cleaner Prod 90:171–182
Saouter E, Biganzoli F, Ceriani L, et al (2018) Environmental footprint: update of life cycle impact assessment methods—ecotoxicity freshwater, human toxicity cancer, and non-cancer, EUR 29495 EN. Publications Office of the European Union, Luxembourg, ISBN 978-92-79-98183-8. JRC114227. https://doi.org/10.2760/611799
Sengupta D, Hawkins TR, Smith RL (2015) Using national inventories for estimating environmental impacts of products from industrial sectors: a case study of ethanol and gasoline. Int J Life Cycle Assess 20:597–607
Seyler C, Hellweg S, Monteil M et al (2004) Life cycle inventory for use of waste solvent as fuel substitute in the cement industry: a multi-input allocation model. Int J Life Cycle Assess 10:120–130
Seyler C, Hofstetter TB, Hungerbuhler K (2005) Life cycle inventory for thermal treatment of waste solvent form chemical industry: a multi-input allocation model. J Cleaner Prod 13:1211–1224

Sharma A, Gulia P (2014) Analysis of big data. IJCSMC 3:56–68
Shibasaki M, Fischer M, Barthel L (2007) Effects on life cycle assessment—scale up of processes. In: Takata S, Umeda Y (eds) Advances in life cycle engineering for sustainable manufacturing businesses. Springer, London, London, p 377
Simon B, Bachtin K, Kilic A et al (2016) Proposal of a framework for scale-up life cycle inventory: a case of nanofibers for lithium iron phosphate cathode applications. Integr Environ Assess Manage 12:465–477
Sinnott RK (2005) Coulson and Richardson's chemical engineering series: volume 6—chemical engineering design, 4th edn. Butterworth-Heinemann, Oxford
Smith RL, Ruiz-Mercado GJ, Meyer DE et al (2017) Coupling computer-aided process simulation and estimations of emissions and land use for rapid life cycle inventory modeling. ACS Sustain Chem Eng 5:3786–3794
Song R, Keller AA, Suh S (2017) Rapid life-cycle impact screening using artificial neural networks. Environ Sci Technol 51:10777–10785
Sonnemann G, Vigon B, Rack M et al (2013) Global guidance principles for life cycle assessment databases: development of training material and other implementation activities on the publication. Int J Life Cycle Assess 18:1169–1172
Strømman A (2001) LCA of hydrogen production from a steam methane reforming plant with CO_2 sequestration and deposition. Paper presented at the 1st Industrial Ecology Conference, Leiden, The Netherlands, 12–14 November 2001
Strogen B, Horvath A (2013) Greenhouse gas emissions from the construction, manufacturing, operation, and maintenance of U.S. distribution infrastructure for petroleum and biofuels. J Infrastruct Syst 19:371–383
Suh S, Huppes G (2002) Missing inventory estimation tool using extended input output analysis. Int J Life Cycle Assess 7:134–140
Suh S, Lenzen M, Treloar GJ et al (2004) System boundary selection in life-cycle inventories using hybrid approaches. Environ Sci Technol 38:657–664
Suh S (2005) Developing a sectoral environmental database for input–output analysis: the comprehensive environmental data archive of the US. Econ Syst Res 17:449–469
Suh S, Huppes G (2005) Methods for life cycle inventory of a product. J Cleaner Prod 13:687–697
Tatari O, Nazzal M, Kucukvar M (2012) Comparative sustainability assessment of warm-mix asphalts: a thermodynamic based hybrid life cycle analysis. Resour Conserv Recycl 58:18–24
Thinkstep (2019) GaBi database. http://www.gabi-software.com/international/databases/gabi-databases/ Accessed 30 May 2019
Treloar G (1997) Extracting embodied energy paths from input-output tables: towards an input-output-based hybrid energy analysis method. Econ Sys Res 9:375–391
US EPA (2006a) TANKS 4.09 Software Program. U.S. EPA, Office of Air Quality Planning and Standards: Research Triangle Park, NC. https://www3.epa.gov/ttnchie1/software/tanks. Accessed 28 Feb 2019
US EPA (2006b) WATER9 3.0 Software Program. U.S. EPA, Office of Air Quality Planning and Standards: Research Triangle Park, NC. https://www3.epa.gov/ttnchie1/software/water/water9_3/index.html. Accessed 28 Feb 2019
US EPA (2010) AP-42 Compilation of Air Pollutant Emission Factors. 5th edn. https://www.epa.gov/air-emissions-factors-and-quantification/ap-42-compilation-air-emissions-factors#5thed. Accessed 28 Feb 2019
US EPA (2016a) Toxicity Forecaster (ToxCast™). https://www.epa.gov/sites/production/files/2016-12/documents/tox_cast_fact_sheet_dec2016.pdf. Accessed 28 Feb 2019
US EPA (2016b) About the TSCA Chemical Substance Inventory. https://www.epa.gov/tsca-inventory/about-tsca-chemical-substance-inventory. Accessed 28 Feb 2019
Van Zelm R, Huijbregts MAJ, van de Meent D (2009) USES-LCA 2.0-a global nested multi-media fate, exposure, and effect model. Int J Life Cycle Assess 14:282–284
Watanabe MDB, Chagas MF, Cavalett O et al (2016) Hybrid input-output life cycle assessment of first- and second-generation ethanol production technologies in Brazil. J Ind Ecol 20:764–774

Wernet G, Hellweg S, Fischer U et al (2008) Molecular-structure-based model of chemical inventories using neural network. Environ Sci Technol 42:6717–6722

Wernet G, Papadokonstantakis S, Hellweg S et al (2009) Bridging data gaps in environmental assessments: modeling impacts of fine and basic chemical production. Green Chem 11:1826–1831

Wernet G, Bauer C, Steubing B et al (2016) The ecoinvent database version 3 (part I): overview and methodology. Int J Life Cycle Assess 21:1218–1230

Wolf K, Mungcharoen T, Yuvaniyama C (2016) 10-year experience with the Thai national LCI database: case study of "refinery products". Int J Life Cycle Assess 22:1760–1770

Williams ED, Weber CL, Hawkins TR (2009) Hybrid framework for managing uncertainty in life cycle inventories. J Ind Ecol 13:928–944

Wright HE, Zhang Q, Mihelcic JR (2008) Integrating economic input–output life cycle assessment with risk assessment for a screening-level analysis. Int J Life Cycle Assess 13:412–420

Yue D, Pandya S, You F (2016) Integrating hybrid life cycle assessment with multiobjective optimization: a modeling framework. Environ Sci Technol 50:1501–1509

Zhang Y, Luo X, Buis JJ et al (2015) LCA-oriented semantic representation for the product life cycle. J Cleaner Prod 86:146–162

Zhou Y, Lee CK, Sharratt P (2017) Bridging the gap from pilot plant experimental records to life cycle inventory. Ind Eng Chem Res 56:10393–10412

Comparison of the LCIA Methods Used for the Evaluation of Chemicals

2

Patrizia Frontera, Beatrice Salieri and Serena Righi

Abstract

Within the four phases of the Life Cycle Assessment, the impact assessment phase involves the evaluation of the environmental impacts of a product/process, object of the study, through its entire lifecycle. In order to categorize and characterize the life cycle impact of the flows to and from environment, several methods have been developed. This variety complicates the comparability of different LCA studies. A paragraph of the chapter is devoted to the USEtox, as recommend method to characterize the toxic impact categories. The purposes of this chapter are to review the different impact assessment methods used in the LCA of chemicals, in order to show that LCIA methods are most appropriate to the environmental assessment of chemical products and to identify the main categories of environmental impact assessed using these methods.

P. Frontera (✉)
Department of Civil, Energy Environments and Materials Engineering,
University Mediterranea of Reggio Calabria, Reggio Calabria, Italy
e-mail: patrizia.frontera@unirc.it

B. Salieri
Swiss Federal Laboratories for Materials Science and Technology (Empa),
St. Gallen, Switzerland

S. Righi
DIFA Department of Physics and Astronomy, CIRSA Inter-Departmental Centre
for Research in Environmental Sciences, University of Bologna, Bologna, Italy

S. Maranghi and C. Brondi (eds.), *Life Cycle Assessment in the Chemical Product Chain*, https://doi.org/10.1007/978-3-030-34424-5_2

2.1 Introduction

The Life Cycle Impact Assessment (LCIA) is the phase of Life Cycle Assessment (LCA) aimed at understanding and evaluating the magnitude and significance of the potential environmental impacts of a product or system. In LCIA, the steps of assignment of inventory are translated in impact categories for the characterisation and for the classification while normalisation and weighting steps are optional elements.

Practically, in the LCIA phase, there is the translation of the inventory results, previously assembled into new information related to the impact of those flows, in order to assess their significance.

For impact assessment of flows resulting from the Life Cycle Inventory (LCI) several methods in the past years have been developed. Following the European Commission Joint Research Center technical notes (EC-JRC 2012) a method is "A set of principle, model and characterization factor" that allows the calculation of the "characterization results for a certain impact category." A methodology collects various characterization methods, which together address the different impact categories.

The choice of suitable impact assessment methods and categories for a reliable LCA is one of the main issues LCA analysts facing regularly. The problem concerning a global consensus on the proper impact assessment for the evaluation of the environmental footprint has been extensively dealt with by the scientific community (Jolliet et al. 2014). This problem is exacerbated in the application of LCA to the chemical sector since the chemical industries produce thousands of new and innovative products every single day.

Therefore, it is essential to identify the suitable impact assessment method, to evaluate the environmental impact along the whole life cycle of chemical products with a global consensus, taking into account the burdens on climate change, human health, biodiversity, water resources, and so on (EC-JRC 2010a). The main goal is to achieve a widely recognized methodological approach useful to increase the comparability of the environmental performances of chemical products and processes and facilitate the decision-making process for the companies and decision makers (ISO 2006a).

In this work, a review of how the LCA studies in the chemical sector are carried out in order to show that LCIA methods are most appropriate to the environmental assessment of chemical products.

2.2 Review of LCIA Methods Used to Assess the Environmental Impact of Chemicals

The "life cycle" of a chemical product refers to the major activities throughout the product's life spanning from its manufacture, use, and maintenance, to its final disposal, including the raw material extraction required for the manufacturing of the

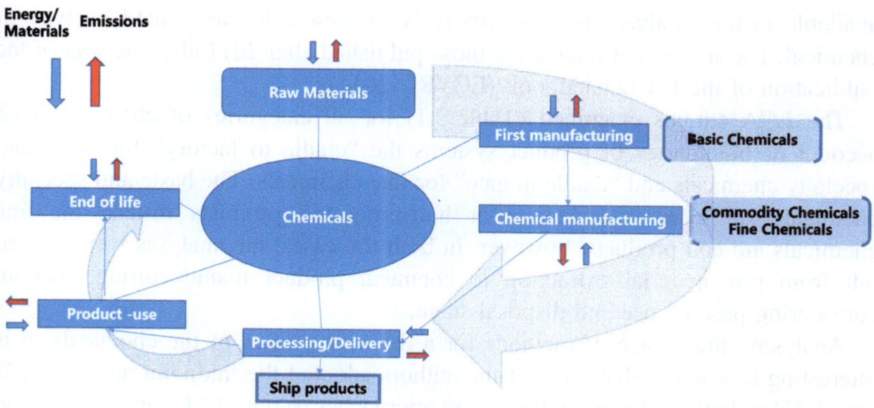

Fig. 2.1 Scheme of the life cycle of chemicals

product. Figure 2.1 reports a typical chemical product life cycle considered in a life cycle assessment (LCA).

In order to carry out the review of LCIA methods used in the chemical industry, the chemicals can be classified into three categories. Commodity or Basic, Specialty. and Fine Chemicals.

- Commodity or Basic Chemicals are produced in huge amounts, are mainly sold within a chemical industry network or to other industries before being transformed into products for the consumer market. As an example, ethanoic acid is sold in order to make esters, the same esters are sold to produce paints for the consumer market.
- Specialty chemicals are chemicals with a high added value that can be added to a number of specific products (such as plastics, coatings, adhesives and inks, cosmetics, and cleaning products) in relatively modest quantities to improve the performance of these products. When specialty chemicals are used for the food market or pharmaceutical industry they are usually named fine chemicals.
- Fine chemicals are complex, single, pure chemical substances, produced in limited quantities within multipurpose plants by multistep batch chemical or biotechnological processes. They are described by exacting specifications, used for further processing within the chemical industry or for the ship products.

This classification was used in order to categorize the different studies analysed, reported in Table 2.1; the literature review was performed using the international scientific database "Science Direct" that covers a considerable range of the important journals in the field of environmental sustainability. The search was carried out in Science Direct using the name of methods and the word "LCA" and separated by the word "AND" (e.g., "International Reference Life Cycle Data System (ILCD)" and "LCA"). The search was done in "all fields" a search type available in the database used; successively the research was refined by type of chemical. The studies considered are those published after 2011 after the year of the publication of the ILCD handbook (EC-JRC 2011).

The LCA studies examined (Table 2.1) for all categories of chemicals take account as boundaries of product systems the "cradle to factory" for basic and specialty chemicals and "cradle to gate" for fine chemicals. The basic and specialty chemicals are often intermediate or half-processed products instead the fine chemicals are end products, however, in both the cases, the analyses were carried out from raw material extraction to chemical product manufacturing, without considering product use and disposal stage.

Analysing the choice of methods for a different classes of the chemicals, it is interesting to observe that most of the authors adopted the midpoint approach (22 out of 31) or both midpoint and endpoint approaches (6 out of 31), on the contrary, only three studies adopted exclusively the endpoint approach. This means that most authors prefer to maintain a detailed perspective, rather than a synthetic one. Moreover, it is worthy of note that most of the studies consider a large number of impact categories rather than one or two more common (global warming and energy demand). The exceptions are: Tallaksen et al. (2015) that applied just IPCC GWP 100a and primary energy; Middlemas et al. (2015) who considered only energy consumption indicators; and Deorsola et al. (2012) who focused on energy demand and greenhouse gas emissions since the process was expected to remarkably reduce energy consumption.

The most adopted midpoint characterization method is CML 2001: thirteen studies have applied it. When the justification for choice of method is supplied, the authors state that CML is often applied in many studies focusing the same subject (Farzard et al. 2017), CML is recognized as one of the most well-established and world-widely used midpoint method (Li et al. 2018) and as the most robust by industry (Bobba et al. 2016). The choice of ILCD 2011 midpoint method v1.03 is explained by Grande et al. (2017) with its ability to support the correct use of characterization factor as recommended in ILCD guidance document (EC-JRC 2010b; Jolliet et al. 2014). Finally, EDIP 2003, chosen only by Moign et al. (2010), was applied since it is a thoroughly documented midpoint approach covering most of the emission-related impacts.

Among the methods that provide both the midpoint and endpoint analysis, ReCiPe is the most selected: seven studies have applied it. ReCiPe is chosen since it allows midpoint and endpoint evaluation (Cespi et al. 2015), considers many different impact categories (Dassisti et al. 2016), is recommended by the European Commission, offers well-detailed documentation and addresses uncertainty in the model (Isola et al. 2017; Chen et al. 2016). A hierarchical perspective provides neither a short nor long-term perspective, as it represents aspects of both (O'Sullivan et al. 2015), so for the assessor ReCiPe2008 method (hierarchic perspective) permitted to enlarge the analysis.

Table 2.1 LCA studies on chemical products. (M) = Midpoint method; (E) = Endpoint method

Chemicals	Methods	Functional unit	References
Basic/commodity chemicals			
Ammonia	CML 2001 (M) Eco-indicator 99 (E)	1 kg ammonia	(Bicer et al. 2016)
Ammonia	IPCC GWP 100a	1 kg N (nitrogen) in the form of anhydrous ammonia	(Tallaksen et al. 2015)
Adipic acid	CML2001 (M) CED	1 kg adipic acid	(Wang et al. 2013)
Adipic acid	CML2001 (M) CED	1 kg of adipic acid	(Aryapratama and Janssen 2017)
Formic acid	IMPACT 2002 + (M)(E)	1 kg of HCOOH	(Shemfe et al. 2018)
1,3-butadiene	CML-IA baseline (M) CED	1 tonne of BD	(Farzad et al. 2017)
Hydrogen	Eco-Indicator 99 (E)	1 kg of produced hydrogen	(Smitkova et al. 2011)
Acrylic acid	CML 2001 (M)	one tonne of acrylic acid	(Petrescu et al. 2016)
Dimethyl ether	ReCiPe 1.08 (M)	1 MJ of net energy (energy content of H_2 and DME) produced	(Schakel et al. 2016)
Oxygen-enriched air (EA)	ReCiPe 1.12 E/A (E)	volume of EA produced in 24 h of plant	(Galli et al. 2018)
Methanol	CML 2001 (M)	1 tonne of methanol	(Li et al. 2018)
Specialty chemicals			
2-methyl tetrahydrofuran	Land footprint, GWP Acidification Eutrophication	1 kg 2-MeTHF	(Khoo et al. 2015)
Limonene	Eco-Indicator 99 (E)	1 L of limonene	(Jahandideh et al. 2017)
Borax and boric acid	CML 2001 (M)	10,000 tonnes of borax and 10,000 tonnes of boric acid	(An and Xue 2014)
Inorganic binders	CML2001 (M)	1 tonne of final product	(Marcelino-Sabada et al. 2017)
C16–C18 triglycerides mixture	ILCD 2011, CML and IMPACT 2002 (M)	1 kg of final product	(Secchi et al. 2016)
Titanium dioxide	CED	1 kg of TiO_2 pigment	(Middlemas et al. 2015)
Penicillin	CML 2 Baseline 2000 V2.03 (M)	1 kg of sodium salt of penicillin V	(Harding et al. 2018)

(continued)

Table 2.1 (continued)

Chemicals	Methods	Functional unit	References
Acrolein	ReCiPe 2008 (v 1.07) (M)	1 kg of acroleine	(Cespi et al. 2015)
Terephthalic acid	IMPACT 2002 + (M)	1 kg of product	(Akunama et al. 2014)
Pesticides	CML-baseline (M)	1 ha treated with the recommended dose	(Margni et al. 2002)
Fine chemicals			
Zinc oxide nanostructures	ReCiPe 2008 (M)	1 m^2/g, the surface area of ZnO nanostructure	(Papadaki et al. 2017)
Molybdenum Sulphide nanoparticles	CED GWP100	1 g of MoS_2 nanoparticles	(Deorsola et al. 2012)
Vanadium mixed acid electrolytes	ReCiPe 2008 (M)	6 L of electrolyte	(Dassisti et al. 2016)
Tungsten disulphide (WS_2)	CML 1992 and ReCiPe 2008 (M)	1 g of nanosized WS_2	(Bobba et al. 2016)
Cadmium sulphide	IMPACT 2002 + (M)	Layers of CdS with an area of 2 cm×2 cm	(Scalbi et al. 2017)
2, 5-furandicarboxylic acid (FDCA),	ReCiPe (M)	1 g of biopolymer derived from FDCA	(Isola et al. 2017)
MOF	ILCD 2011 (M)	1 kg of MOF powder	(Grande et al. 2017)
Titanium dioxide	BEES	0.046 ml of the product of 1m^2 coating	(Babaizadeh and Hassan 2013)
Lead zirconate titanate	Eco-Indicator 99 (M)	1 kg of product	(Ibn-Mohammeda et al. 2018)
Yttria-stabilized zirconia	EDIP 2003 (M)	One micrometre-thick layer of YSZ on 1m^2 surface	(Moign et al. 2010)

IMPACT 2002 + has been applied in four studies, just Scalbi et al. (2017) motivated the choice arguing that it includes some impact categories which are particularly significant for the goal and scope of their study and at the same time allows an overall environmental analysis.

The endpoint method Eco-Indicator 99 has been chosen by four authors. Smitkova et al. (2011) selected it because grouping into tree damage categories was considered to be an ideal mix of impacts for their goals. The other authors do not provide information about the choice.

Identification of impact categories selected by authors that have followed the midpoint approach (solely or combined with endpoint) is summarised in Fig. 2.2. The figure illustrates the application percentage of each impact category in the case studies review. As expected, since its prominence in the environmental protection filed, climate change is the impact category with the highest frequency of use for all

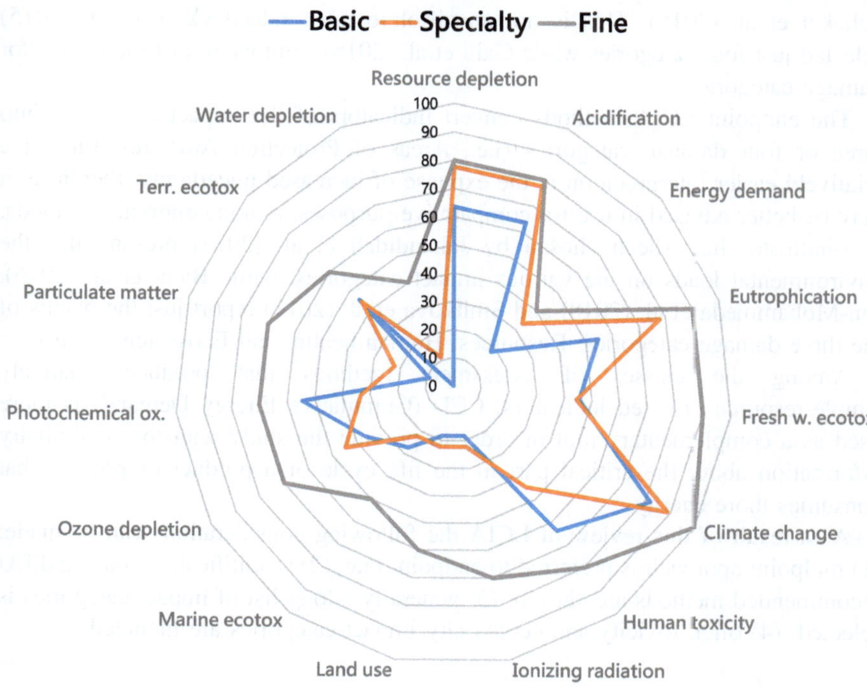

Fig. 2.2 Frequency of different impact categories used in reviewed cases studied

types of chemicals-basic, specialty or fine. The only ones who do not select climate change were Middlemas et al. (2015) since they focused on energy demand. Very frequently used impact categories are also acidification, eutrophication, resource depletion, and human toxicity. The latter is more considered for the fine chemicals (in 8 papers out of 11) and basic (in 7 out of 9) than the specialty (in 4 out of 8). This may be explained by the different typology of substances of products belonging to different classes. Many other impact categories are applied in a gradually decreasing way (aquatic and terrestrial ecotoxicity, photochemical oxidation, ozone depletion, ionizing radiation, etc.). It is noteworthy that some authors have proposed quite new and original impact categories like land footprint (Khoo et al. 2015), indoor air quality and criteria air pollution (Babaizadeh and Hassan 2013), malodours air and materials utilisation (Ibn-Mohammeda et al. 2018).

Using the midpoint LCIA methods, such as CML 2001, the environmental interventions are represented as a set of different impact categories, so can be difficult to interpret the midpoint results, for this in the most part of cases examined the commented impact categories are reduced (Bicer et al. 2016).

Users of ReCiPe generally prefer to report midpoint and endpoint approach results and consequently a long number of impact categories are presented by Bobba et al. (2016), Dassisti et al. (2016), Isola et al. (2017), Papadaki et al. (2017),

Schakel et al. (2016). This is not an absolute rule, indeed Cespi et al. (2015) selected just four categories while Galli et al. (2018) summarize just the results for damage categories.

The endpoint LCIA methods convert indicators of the impact categories into three or four damage categories (i.e., Areas of Protection AoP) resulting in a relatively easier interpretation at the expense of increased uncertainty, therefore, it may be better advised in use for comparative purposes. Among endpoint methods, Eco-indicator have been chosen by Jahandideh et al. (2017) present also the environmental loads on the various impact categories, while Bicer et al. (2016), Ibn-Mohammeda et al. (2018) and Smitkova et al. (2011) report just the results of the three damage categories: Resources, Human health, and Ecosystem quality.

Among the classes of assessment methods that produce relatively simple-resource oriented indicators, CED (Cumulative Energy Demand) is more used as a complementary tool in order to provide the study with the preliminary information about the critical part in the life cycle of a product or process that consumes more energy.

As a result of this review of LCIA the following considerations can be made: (1) midpoint approach is preferred to endpoint one; (2) scientifically robust and EU recommended methods are chosen; (3) generally a long list of impact categories is selected; (4) often toxicity and ecotoxicity impact categories are included.

2.3 Critical Issues in the Choice of the Methods

The choice of LCIA methods and impact categories is an essential decision along with the LCA case study, despite this, very little guidance is available in literature to support LCA practitioners (Haes et al. 1999; Rosenbaum 2018).

The difficulty in choosing is due to the scientific complexity of the models related to the various impact categories and to the "fake similarity" between LCIA methods with similar impact categories that "hide" different scientific models. Indeed, although different LCIA methods partially include the same impact categories, they do not use the same scientific models. For example, CML2002 calculates Photochemical Ozone Creation Potential (POCP) applying a Lagrangian type model (Derwent et al. 1998) while EDIP2003 uses a Eulerian model (Heyes et al. 1996).

The ISO 14044 (ISO 2006b) established recommendation and requirements for the choice of impact categories, category indicators, characterization models within the LCA study. As requirements, the ISO 14044 established, i.e., that the choice of impact categories shall be consistent with the goal and the study and comprehensiveness of the environmental issues related to the product system under investigation. Also, recommendation as such the impact method, impact categories should be scientifically and technically valid and internationally accepted. For LCA practitioners, Rosenbaum 2018 provided useful recommendations and selection criteria.

Numerosity of Impact Assessment Methods Several LCIA methods have emerged since the early 1990s. This situation has created some confusion since the results of the LCIA vary according to the LCIA method selected (European Commission Joint Research Centre Institute for Environment and Sustainability 2010b).

Furthermore, it is difficult to compare the results of a midpoint approach with those of an endpoint one (Dreyer et al. 2003). The international community has, therefore, committed itself to standardize LCIA methods to obtain more consensus. The intention to join a standardization is evidenced by the development of combined approach methods such as ReCiPe (Hauschild et al. 2015) and IMPACT 2002 + (Jolliet et al. 2003) and consensus methods such as USEtox® and the recommendations issued by the Joint Research Centre (European Commission Joint Research Centre Institute for Environment and Sustainability 2012). In spite of these efforts, it is unlikely that the LCA methodology will have unique methods because it is essential to maintain the diversity of the different specific contexts of the studied systems.

Number of Substances Considered in the LCA Methods The scientific models used to evaluate a given ambient phenomenon (e.g., acidification, eutrophication, and ozone depletion) should study each substance that could contribute to this phenomenon, in order to be able to shape the fate of the substances in question. As an example, it is now recognized on a horizon of 100 years, that methane has a greenhouse effect 28 times greater than CO_2 (Myhre et al. 2013), but the research was necessary to reach this conclusion.

Currently, the substances included in characterization models remain relatively limited and just few LCIA methods cover more than 1000 substances (Hauschild et al. 2008). This number is negligible considering that EINECS (European Inventory of Existing Commercial Chemical Substances) lists more than 100,000 substances currently placed on the market many of which with effects unknown.

Therefore, if the inventory data contains, for example, 2000 different substances, according to the selected LCIA method, a certain part of the substances will not be considered and their effects will not be assessed. Obviously, this condition makes complicate comparison among different methods. It is necessary to highlight that an LCIA method that has few substances which can, however, include substances with great potential impact.

Spatial Differentiation of Impacts Environmental impacts have different areas of influence. There are global, continental, regional, and local impacts. Expect global impacts (climate changes and depletion of the ozone layer), the potential environmental impact caused by the emission of a substance depends, among other things, on the receiving environment. For example, the presence of alkaline salts in soils or waters in a given area influences the phenomenon of acidification with respect to an area lacking these salts. In the same way, a high-density area will have higher exposure risk to the release of a toxic substance compared to an area with few inhabitants per km^2.

Therefore, the evaluation of the environmental impacts of a substance should consider also the area where it is emitted and its fate in the environment. LCIA methods should be able to take into account the territorial disparities in order to guarantee a fair distribution of the potential impacts on the environment. Many developments have been made in this regard. Indeed, many LCIA methods have been developed for Europe, Japan, North America, and some of these consider regional, continental, and global differentiation. However, it is difficult for the methods to take into account the differences within the same region.

2.4 An Overview of Selected Methods for Impact Categories of Interest for Environmental Evaluation of Chemicals

There are numerous impact assessment methods available to LCA analysis that can be categorized following the approaches described below: (i) midpoint approach, (ii) endpoint approach, (iii) combined midpoint and endpoint approaches. As defined in (ILCD 2011), the midpoint and endpoint approaches are characterization models that provide indicators at different levels.

A recent overview of existing LCIA methods have been carried out by Rosenbuam (Rosenbaum 2018) and before by Hauschild et al. (2013); however, the ReCiPe 2016 update was not taken into account, as not published yet. Therefore, in this paragraph, we will compare ReCiPe 2016, developed by the partnership between four leading institutions (Goedkoop et al. 2008, 2009), with the Institute for Environmental Science (CML) impact assessment tool (more recent updates as CML-IA baseline and not baseline) developed at the University of Lieden in the Netherlands. So, an endpoint and a midpoint method will be compared; the CML method has been analysed because it is the most used in the LCA studies considered in this review. In order to carry out this comparative evaluation, seven categories of impact were selected, since the objective of this chapter is also to be a support tool for the choice of LCIA in the study of chemicals. The selection of impact categories was made to meet specific features of this sector, but also to address concerns about environmental issues (laws, policies, international agreements, etc.). Aside from the toxicity, which will be studied further on, the categories considered are climate change, ozone depletion, photochemical smog formation, acidification, aquatic eutrophication, land use, resource consumption.

A first significant difference between the two methods is the number of substances taken into consideration for the evaluation of the impacts of the different categories, as shown in Fig. 2.3.

Climate Change This category of impact is one among the most emphasized by the media. Obviously, it has significant relevance in the study of LCAs of chemicals, especially for those involving large production volumes and requiring

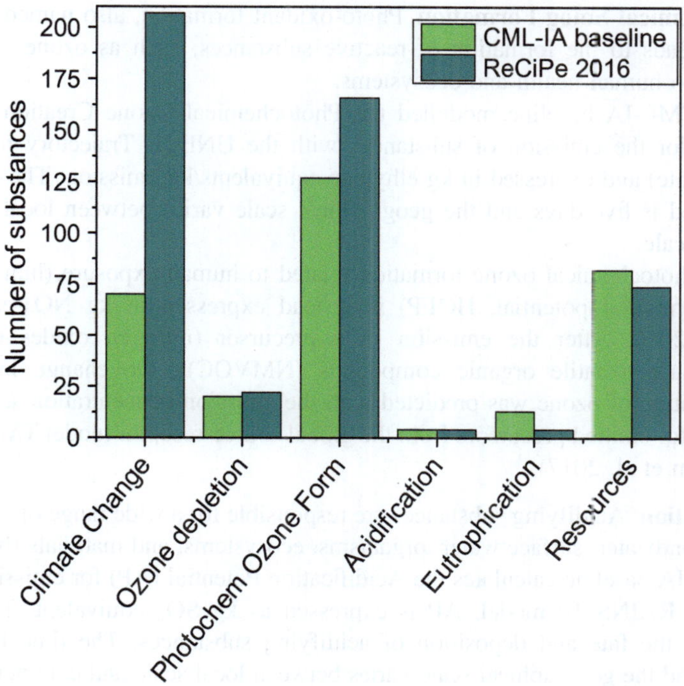

Fig. 2.3 Number of substances included in the CML-IA baseline and ReCiPe 2016 for selected impact categories

numerous unit operations (i.e., adsorption, distillation, and purification) for their production, requiring high expenditure of energy.

Both the CML-IA baseline method and the ReCiPe 2016 one use as characterization model the one developed by the IPCC Intergovernmental Panel on Climate Change (IPCC) and the factors are expressed as Global Warming Potential GWP. For both methods, the geographic scope of this method is at a global scale.

Ozone Depletion A big fraction of UV-B radiation is able to reach the earth's surface due to the stratospheric ozone depletion. Consequently, the effects upon human and animal health, terrestrial, and aquatic ecosystem can be damaged.

The definition of ozone depletion potential of different gases (kg CFC-11 equivalent/kg emission), developed by the World Meteorological Organization (WMO), is the characterization model used.

In ReCiPe 2016, for endpoint level, three-time horizons have been implemented: 20 years (Individualist), 100 years (Hierarchist), and infinite (Egalitarian). The geographic scope of this indicator is at global scale.

Photochemical Smog Formation Photo-oxidant formation, also named "summer smog", leads to the formation of reactive substances, such as ozone, which are harmful to human health and ecosystems.

The CML-IA baseline modelled the Photochemical Ozone Creation Potential (POCP) for the emission of substances with the UNECE Trajectory model (including fate) and expressed in kg ethylene equivalents/kg emission. The time span considered is five days and the geographical scale varies between local and continental scale.

The photochemical ozone formation related to human exposure (human health ozone formation potential, HOFP) is instead expressed in kg NO*x*-eq for the ReCiPe 2016. After the emission of a precursor (nitrogen oxides (NOx) or non-methane volatile organic compounds (NMVOC)), the change in ambient concentration of ozone was predicted with the emission-concentration sensitivities matrices for emitted precursors from the global source-receptor model TM5-FASST (van Zelm et al. 2017).

Acidification Acidifying substances are responsible for a wide range of impacts on soil, groundwater, surface water, organisms, ecosystems, and materials (buildings).

CML-IA baseline calculates the Acidification Potential (AP) for emissions to air by using RAINS 10 model. AP is expressed as kg SO_2 equivalent. The model describes the fate and deposition of acidifying substances. The time horizon is infinite and the geographical scale varies between local scale and continental scale.

As for the ReCiPe 2016, the fate of a pollutant in the atmosphere and the soil as calculated by (Roy et al. 2014) was considered for the midpoint characterization factors of acidifying emissions. Acidification potential (AP) is expressed in kg SO_2-equivalents. The GEOS-Chem model (Roy et al. 2012a) was used to calculate the changes in acid deposition, following changes in air emission of NO*x*, NH_3, and SO_2. The geochemical steady state model PROFILE (Roy et al. 2012b) was then considered to evaluate the change in acidity in the soil due to a change in acid deposition.

Eutrophication Eutrophication or nitrification is an enrichment of water, air, and soil by nutrient salts that causes changes to the ecosystem, with all consequent impacts.

For the CML-IA baseline, Nutrification Potential (NP) is expressed as kg PO_4 equivalents per kg emission and it is based on the stoichiometric procedure.

Fate and exposure are not included, the considered time span is eternal, and the geographical scale varies between local and continental scale.

The fate of phosphorus forms the basis of the midpoint characterization factors for freshwater eutrophication, for the ReCiPe 2016. Freshwater eutrophication potentials (FEP) are expressed in kg P to freshwater-equivalents. Global fate factors for phosphorus emissions to freshwater were taken from (Helmes et al. 2012).

Resources Consumption For the CML-IA baseline, the resources consumption impact category indicator is related to the extraction of minerals and fossil fuels due to inputs in the system. The Abiotic Depletion Factor (ADF) considers concentration reserves and rate of deaccumulation for each extraction of minerals and fossil fuels (kg antimony equivalents/kg extraction). The geographic scope of this indicator is at global scale.

For the ReCiPe 2016, Surplus Ore Potential (SOP), expressed as kg Cu-eq, is considered as the midpoint characterization factor for mineral resource scarcity. The primary extraction of a mineral resource leads to an overall decrease in ore grade (the concentration of that resource in ores worldwide) which in turn will increase the amount of ore produced per kilogram of mineral resource extracted. The SOP reports the average extra quantity of ore produced in the future due to the extraction of a mineral resource considering all future production of that mineral resource. The ratio between the higher heating value of a fossil resource and the energy content of crude oil is defined as the midpoint indicator for fossil resource use, determined as the Fossil Fuel Potential (FFP in kg oil-eq).

Land Use Although this category of impact is considered with attention in the LCA studies relating to agri-food systems, it reveals particular importance for all the chemicals that are naturally derived, such as biofuels.

In the CML-IA baseline method, the land use category is not considered.

For the ReCiPe 2016 the characterization factors of midpoints of agricultural land occupation, urban land occupation, and natural land transformation, evaluated in m^2 yr annual crop equivalents, refers to the relative species reduction due to a specific land use type such as annual crops, permanent crops, mosaic agriculture, forestry, urban land, and pasture. Relative species loss was determined by comparing field data on local species richness in specific types of natural and human-made land covers (De Baan et al. 2013). Based on average recovery times from (Curran et al. 2014), passive recovery towards a (semi-) natural old-growth habitat was assumed for land conversion.

2.5 Assessment of Toxicity in Life Cycle Impact Assessment

The evaluation of the toxicity impacts related to a chemical emission into the environment requires characterization models, and factors, accounting for chemical fate in the environment, exposure, and differences in toxicological response (EC-JRC 2010a).

Over years, several models for the toxic impact categories have been developed on the basis of different algorithms and assumptions that can lead to different results (Hauschild et al. 2011) and thus, it might contribute to the lack of addressing toxic impacts in LCA studies. In 2003, to overcome the large variation among the available methods, the UNEP/SETAC Life Cycle Initiative established a Taskforce on Toxic Impact to provide guidance for the assessing of toxic-related impact categories (Westh et al. 2015). The task Force recognized as the existing toxicity assessment models were covering a limited number of substances and that the scope, principles, and results varied substantially among them.

Thus, an international model consensus process was initiated that resulted in the development of the USEtox™ model. (Rosenbaum et al. 2008; Hauschild et al. 2008). The USEtox model is a scientific consensus model developed for the characterization of the human and ecotoxicological impacts of chemicals.

The model provides characterization factors (CF) for organic (3077 substances) and inorganic substance (27 substance, all cationic metals) for three toxic impact categories, namely human cancer toxicity, human non-cancer toxicity, and freshwater aquatic ecotoxicity at midpoint and endpoint level (Hertwich et al. 2001; Kounina et al. 2014).

The USEtox model is a multimedia fate-exposure-toxicity model and applies a matrix-algebra framework. The unit of the CFs for freshwater aquatic ecotoxicity has been established as potentially affected fraction (PAF) of aquatic species integrated over the time (days) and the exposed water volume (m^3), whereas, for the human toxicity the unit as cumulative case of cancer/non-cancer health outcomes per kg emitted is used. The human health and freshwater ecotoxicity impacts are expressed using a Comparative Toxic Unit (CTU) approach, to support the comparative nature of the characterization factors (Fantke et al. 2017).

The CFs are calculated as (2.1):

$$CF = FF \cdot XF \cdot EF \tag{2.1}$$

where, FF—the fate factor—quantifying how the chemical emission is dispersed in the environment, XF—exposure factor—quantifying the human intake and the ecological system contact with the media, the effect factor (EF) quantifying the toxicity of the chemical to humans and freshwater aquatic species (Fantke et al. 2017).

The fate of the substance into the environment is assessed by nested multimedia fate model and solving a set of mass balance equations at the steady state describing transformation process (i.e. biological, physical, and chemical degradation, hydrolysis, oxidation, etc.) removal process (i.e., leaching, burial) and inter-compartment transfer and advective processes. The fate model accounts for the environmental compartments of indoor air, urban air, agricultural soil, natural soil, freshwater, coastal marine, and oceans. Four spatial scales are considered: indoor, urban, continental, and global. The environmental compartments are represented as homogeneous and well-mixed boxes. The fate model calculates the FFs (day) representing the persistence and distribution of a chemical emission into the environment. For the impact category of human toxicity, the XFs (day^{-1}) represents the rate at which a pollutant is transferred from a receiving environmental compartment into the human population. It accounts for several exposure pathways: inhalation or ingestion of drinking water.

For the impact category of freshwater aquatic ecotoxicity, the XFs (dimensionless) represents the fraction of the dissolved chemical in freshwater; the effect factor EFs (PAF m^3 kg^{-1}) represent the potentially affected fraction of species PAF over a volume per of exposed substance (m^3 kg^{-1}). The EFs (cases kg^{-1}intake) for human toxicity represents the probability of adverse effect (carcinogenic or no carcinogenic effects) related to the amount of the chemical taken in by the population via inhalation or ingestion (Rosenbaum et al. 2008).

USEtox model provides, also, CFs at endpoint level by accounting for damage factor expressed as disability adjusted life years (DALY) for human health impact and potentially disappeared fraction of species (PDF) for freshwater aquatic toxicity.

Since its first release (version 1.01) a series of additional substances, exposure pathways, regionalized landscape data, and features have been implemented in USEtox. Notably, in version 2.0 residential and indoor environmental has been added allowing to calculate CF for potential human toxicity impacts of pollutants emitted indoor. Also, the exposure to pesticide residues via food crop consumption and new CFs for 14 cationic metal in freshwater were included.

In USEtox, "indicative" CFs are specified reflecting the level of reliability of the calculation related to the high uncertainty of addressing environmental fate and human exposure (Fankte et al. 2017). Thus, their impact scores need to be interpreted cautiously. The following groups of organic substances have been classified as "indicative": organometallic substances, amphiphilic, and dissociating substances. Also, all the CFs for metals are flagged as "indicative" due to the high uncertainty associated with estimates of fate, exposure, and effects for this substance group.

USEtox characterization factors can vary more than 12 orders of magnitude across the chemical and within a factor of 100–1,000 for CFs for human health and 10–100 for CFs for freshwater ecotoxicity. This is not surprising since it is due to the vast difference across chemicals (i.e., in amount produced, distribution process in the environment, differences in species sensitivity, etc.) (Rosenbaum et al. 2008). This has discouraged the LCA practitioner to include the toxicity impact categories in LCA studies. However, it should be a remark that toxicity models need to account for a vast amount of chemicals and the inherent variability among those substances will always results with uncertainty. So far, by considering the "inherent" uncertainty related to toxic impact categories, those impacts should not be "a priori" excluded from LCA studies, but rather the toxicity impact score should be interpreted considering their uncertainty.

USEtox has been recommended by the European Commission, in the context of the ILCD handbook and as recommended model at midpoint in the context of Product Environmental Footprint (PEF). Also, USEtox characterization factors have been implemented in LCA software (e.g., GaBi, SimaPro, OpenLCA, Quantis Suite) and some LCIA methods (e.g., ILCD LCIA, Impact World+, TRACI 2).

Conclusions

In this chapter, a review has been carried out, examining the various impact assessment methodologies used in the LCA of chemicals, as well as identifying the main categories of environmental impact assessed with such methods. Notably, most of the studies reviewed consider a large number of impact categories rather than the most common ones, e.g., global warming and energy demand. The application as percentage of every impact category has been quantified confirming that, as expected, the climate change is the impact category used with the highest frequency.

The most critical issues in the choice of methodologies have been singled out. This is a matter difficult to be faced, because of the scientific complexity of the characterization models related to the different impact categories, the numerosity of impact assessment methods, the number of substances considered in LCA.

Moreover, an overview of selected methods for impact categories of interest for environmental assessment of chemicals (including climate change ozone depletion, photochemical ozone formation, acidification, eutrophication, resources) is reported. Two LCIA methods (ReCiPe and CML) have been compared. A significant difference between the two methodologies is apparent.

Finally, for the evaluation of toxic impact categories in LCA, the USEtox model is recognized as the most advanced model currently available and is recommended by several organizations and authorities.

References

Akanuma Y, Selke S, Auras R (2014) A preliminary LCA case study: comparison of different pathways to produce purified terephthalic acid suitable for synthesis of 100% bio-based PET. Int J Life Cycle Assess 19(6):1238–1246

An J, Xue X (2014) Life cycle environmental impact assessment of borax and boric acid production in China. J Clean Prod 66:121–127

Aryapratama R, Janssen M (2017) Prospective life cycle assessment of bio-based adipic acid production from forest residues. J Clean Prod 164:434–443

Babaizadeh H, Hassan M (2013) Life cycle assessment of nanosized titanium dioxide coating on residential windows. Constr Build Mater 40:314–332

Bicer Y, Dincer I, Zamfirescu C et al (2016) Comparative life cycle assessment of various ammonia production methods. J Clean Prod 135:1379–1395

Bobba S, Deorsola FA, Blengini GA et al (2016) LCA of tungsten disulphide (WS2) nano-particles synthesis: state of art and from-cradle-to-gate LCA. J Clean Prod 139:1478–1484

Cespi D, Passarini F, Mastragostino G et al (2015) Glycerol as feed-stock in the synthesis of chemicals: a life cycle analysis for acrolein production. Green Chem 17:343–355

Chen L, Pelton REO, Smith TM (2016) Comparative life cycle assessment of fossil and bio-based polyethylene terephthalate (PET) bottles. J Cleaner Prod 137:667–676

Curran M, Hellweg S, Beck J (2014) Is there any empirical support for biodiversity offset policy? Ecol Appl 24(4):617–632

Dassisti M, Cozzolino G, Chimienti M et al (2016) Sustainability of vanadium redox-flow batteries: Benchmarking electrolyte synthesis procedures. Int J Hydrog Energy 41(37):16477–16488

De Baan L, Alkemade R, Köllner T (2013) Land use impacts on biodiversity in LCA: a global approach. Int J Life Cycle Assess 18(6):1216–1230

Deorsola FA, Russo N, Blengini GA et al (2012) Synthesis, characterization and environmental assessment of nanosized MoS2 particles for lubricants applications. Chem Eng J 195–196:1–6

Derwent RG, Jenkin ME, Saunders SM et al (1998) Photochemical ozone creation potentials for organic compounds in northwest Europe calculated with a master chemical mechanism. Atmos Environ 32(14/15):2429–2441

Dreyer LC, Niemann AL, Hauschild MZ (2003) Comparison of three different LCIA methods: EDIP97, CML2001 and Eco-indicator 99. Does it matter which one you choose? Int J Life Cycle Assess 8(4):191–200

European Commission, Joint Research Centre, Institute for Environment and Sustainability (2010a) International reference life cycle data system (ILCD) handbook-frameworks and requirements for life cycle impact assessment models and indicators, 1st edn

European Commission, Joint Research Centre Institute for Environment and Sustainability (2010b) International reference life cycle data system (ILCD) handbook-analysis of existing environmental impact assessment methodologies for use in life cycle assessment (LCA), 1st edn

European Commission, Joint Research Centre, Institute for Environment and Sustainability (2011) International reference life cycle data system (ILCD) handbook-recommendations for life cycle impact assessment in the European context, 1st edn

European Commission, Joint Research Centre, Institute for Environment and Sustainability (2012) Characterisation factors of the ILCD recommended life cycle impact assessment methods. Database and Supporting Information. First edition, Luxembourg

Fantke P, Bijster M, Guignard C et al (2017) USEtox® 2.0. Documentation Version 1

Farzad S, Mandegari MA, Görgens JF (2017) Integrated techno-economic and environmental analysis of butadiene production from bio-mass. Bioresour Technol 239:37–48

Galli F, Pirola C, Previtali D et al (2018) Eco design LCA of an innovative lab scale plant for the production of oxygen-enriched air. Comparison between economic and environmental assessment. J Clean Prod 171:147–152

Goedkoop HR, Huijbregts MAJ, De Schryver A et al (2008) A life cycle impact assessment method which comprises harmonised category indicators at the midpoint and the endpoint level. Version 1.08

Goedkoop M, Heijungs R, Huijbregts MAJ et al (2009) ReCiPe 2008: a life cycle impact assessment method which comprises harmonised category indicators at the midpoint and endpoint levels. First edition, Report i: Characterization. The Netherlands: Ruimte en Milieu, Ministerie van Volkshuisvesting, Ruimtelijke Ordening en Milieubeheer

Grande CA, Blom R, Spjelkavik A et al (2017) Life-cycle assessment as a tool for eco-design of metal-organic frameworks (MOFs). Sustain Mat and Technol 14:11–18

Haes HAU, Jolliet O, Finnveden G et al (1999) Best available practice regarding impact categories and category indicators in life cycle impact assessment. Int J Life Cycle Assess 4(2):66–74

Harding KG, Dennis JS, Harrison STL (2018) Generic flowsheeting approach to generating first estimate material and energy balance data for life cycle assessment (LCA) of penicillin V production. Sust Prod Consum 15:89–95

Hauschild M, Goedkoop M, Guinée JB et al (2013) Identifying best existing practice for characterization modeling in life cycle impact assessment. Int J Life Cycle Assess 18:683–697

Hauschild MZ, Huijbregts MAJ (eds) (2015) Life cycle impact assessment. In: Klöpffer W, Curran M (eds) LCA compendium—the complete world of life cycle assessment. Springer, Dordrecht

Hauschild MZ, Huijbregts MAJ, Jolliet O et al (2008) Building a model based on scientific consensus for life cycle impact assessment of chemicals: the search for harmony and parsimony. Environ Sci Technol 42:7032–7037

Hauschild MZ, Jolliet O, Huijbregts MAJ (2011) A bright future for addressing chemical emissions in life cycle assessment. Int J Life Cycle Assess 18:697–700

Heyes C, Schöpp W, Amann M, Unger S (1996) A reduced-form model to predict long-term ozone concentrations in Europe. Interim report WP-96- 12/December, IIASA, Vienna

Helmes RJK, Huijbregts MAJ, Henderson AD et al (2012) Spatially explicit fate factors of phosphorous emissions to fresh water at the global scale. Int J Life Cycle Assess 17(5):646–654

Hertwich E, Matales SF, Pease WS et al (2001) Human toxicity potentials for life-cycle assessment and toxics release inventory risk screening. Environ Technol Chem 20:928–939

Ibn-Mohammed T, Reaneyc IM, Koha SCL et al (2018) Life cycle assessment and environmental profile evaluation of lead-free piezoelectrics in comparison with lead zirconate titanate. J Eur Ceram Soc 38(15):4922–4938

ISO (2006a) Environmental management-life cycle assessment-principles and framework. International Standardization Organization (ISO), European Standard EN ISO, Geneva, p 14040

ISO (2006b) Environmental management-life cycle assessment-requirements and guidelines. International Standardisation Organisation (ISO), European Standard EN ISO, Geneva, p 14044

Isola C, Sieverding HL, Raghunathan R et al (2017) Life cycle assessment of photodegradable polymeric material derived from renewable bioresources. J Clean Prod 142:2395–2944

Jahandideh A, Johnson TJ, Esmaeili N et al (2017) Life cycle analysis of a large-scale limonene production facility utilizing filamentous N2-fixing cyanobacteria. Algal Res 23:1–11

Jolliet O, Margni M, Charles R et al (2003) IMPACT 2002 + : a new life cycle impact assessment methodology. Int J Life Cycle Assess 8:324

Jolliet O, Frischknecht R, Bare J et al (2014) Global guidance on environmental life cycle impact assessment indicators: findings of the scoping phase. Int J Life Cycle Assess 19:962–967

Khoo HH, Wong LL, Tan J et al (2015) Synthesis of 2-methyl tetrahydrofuran from various lignocellulosic feedstocks: Sustainability assessment via LCA. Resour Conserv Recycl 95:174–182

Kounina A, Margni M, Shaked S et al (2014) Spatial analysis of toxic emissions in LCA: a sub-continental nested USEtox model with freshwater archetypes. Environ Int 69:67–89

Li J, Ma X, Liu H, Zhang X (2018) Life cycle assessment and economic analysis of methanol production from coke oven gas compared with coal and natural gas routes. J Clean Prod 185:299–308

Marcelino-Sadaba S, Kinuthi J, Oti J et al (2017) Challenges in life cycle assessment (LCA) of stabilised clay-based construction materials. Appl Clay Sci 144:121–130

Margni M, Rossier D, Crettaz P et al (2002) Life cycle impact assessment of pesticides on human health and ecosystems. Agr Ecosyst Environ 93:379–392

Middlemas S, Fang ZZ, Fan P (2015) Life cycle assessment comparison of emerging and traditional titanium dioxide manufacturing processes. J Clean Prod 89:137–147

Moign A, Vardelle A, Themelis NJ et al (2010) Life cycle assessment of using powder and liquid precursors in plasma spraying: The case of yttria-stabilized zirconia. Surf Coat Tech 205 (2):668–673

Myhre G, Shindell D, Bréon FM et al (2013). Anthropogenic and natural radiative forcing. In: Stocker TF, Qin D, Plattner GK, Tignor M, Allen SK, Boschung J, Nauels A, Xia Y, Bex V, Midgley PM (eds) Climate change 2013: the physical science basis. Contribution of working group i to the fifth assessment report of the intergovernmental panel on climate change. Cambridge University Press, Cambridge, United Kingdom and New York, NY, USA, p 658

O'Sullivan AD, Wicke D, Hengen TJ et al (2015) Life cycle assessment modelling of stormwater treatment systems. J Environ Manage 149:236–244

Papadaki D, Foteinis S, Mhlongo GH et al (2017) Life cycle assessment of facile microwave-assisted zinc oxide (ZnO) nanostructures. Sci Total Environ 586:566–575

Petrescu L, Fermeglia M, Cormos CC (2016) Life cycle analysis applied to acrylic acid production process with different fuels for steam generation. J Clean Prod 133:294–303

Roy PO, Azevedo LB, Margni M et al (2014) Characterization factors for terrestrial acidification at the global scale: a systematic analysis of spatial variability and uncertainty. Sci Total Environ 500:270–276

Roy PO, Deschenes L, Margni M (2012a) Life cycle impact assessment of terrestrial acidification: modelling spatially explicit soil sensitivity at the global scale. Environ Sci Technol 46 (15):8270–8278

Roy PO, Huijbregts M, Deschenes L et al (2012b) Spatially-differentiated atmospheric source-receptor relationships for nitrogen oxides, sulfur oxides and ammonia emissions at the global scale for life cycle impact assessment. Atmos Environ 62:74–81

Rosenbaum RK, Bachmann TMK, Gold LS et al (2008) USEtox-The UNEP/SETAC-consensus model: recommended characterisation factors for human toxicity and freshwater ecotoxicity in Life Cycle Impact Assessment. Int J Life Cycle Assess 13:532–546

Rosenbaum RK (2018) Selection of impact categories, category indicators and characterisation models in goal and scope definition. In: Curran MA (ed) LCA compendium-the complete world of life cycle assessment-goal and scope definition in life cycle assessment. Springer, Dordrecht, pp 63–122

Scalbi S, Fantin V, Antolini F (2017) Environmental assessment of new technologies: Production of a Quantum Dots-Light Emitting Diode. J Clean Prod 142(4):3702–3718

Schakel W, Oreggioni G, Singh B et al (2016) Assessing the techno-environmental performance of CO2 utilization via dry reforming of methane for the production of dimethyl ether. J CO2 Util 16:138–149

Secchi M, Castellani V, Collina E et al (2016) Assessing eco-innovations in green chemistry: life cycle as-sessment (LCA) of a cosmetic product with a bio-based ingredient. J Clean Prod 129:269–281

Shemfe M, Gadkari S, Yu E et al (2018) Life cycle, techno-economic and dynamic simulation assessment of bioelectrochemical systems: a case of formic acid synthesis. Bioresour Technol 255:39–49

Smitkova M, Janíček F, Riccardi J (2011) Life cycle analysis of processes for hydrogen production. Int J Hydrog Energy 36(13):7844–7851

Tallaksen J, Bauer F, Hulteberg C et al (2015) Nitrogen fertilizers manufactured using wind power: greenhouse gas and energy balance of community-scale ammonia production. J Clean Prod 107:626–635

Van Zelm R, Preiss P, van Goethem T et al (2017) Regionalized life cycle impact assessment of air pollution on the global scale: damage to human health and vegetation. Atmos Environ 134:129–137

Wang Q, Vural Gürsel I, Shang M et al (2013) Life cycle assessment for the direct synthesis of adipic acid in microreactors and benchmarking to the commercial process. Chem Eng J 234:300–311

Westh TB, Hauschild MZ, Birkved M et al (2015) The USEtox story: a survey of model developer visions and user requirements. Int J Life Cycle Assess 20:299–310

LCA Integration Within Sustainability Metrics for Chemical Companies

3

Daniele Cespi, Fabrizio Passarini, Esmeralda Neri, Raffaele Cucciniello and Fabrizio Cavani

Abstract

The application of a life-cycle perspective within the industrial sector may help companies in supporting all the planning activities aimed to promote new business opportunities. The usage of LCA is a common practice in corporates working in the field of chemistry. The development and production of fine/bulk chemicals, pharmaceuticals, plastics, personal care products, etc., may be supported by LCA and green metrics. The development of a lower impact and safer chemical industry is encouraged by the adoption of the Green Chemistry principles. Among these the usage of renewables sources of building blocks is one of the most investigated principle. However, the use of biomass as starting precursors needs to be assessed through LCA before considering a bio-based route greener than the traditional fossil pathway. Recently, the social sphere of sustainability has increased its importance also in the chemical industry, because many chemicals could have social repercussions as a consequence of their adoption. The implementation of S-LCA strategies may help enterprises to configure repercussions of their activities in this sense, by achieving SDGs described in the Agenda 2030. This chapter is intended to drive readers through such issues by stimulating their sensibility towards sustainability within the chemical industry.

D. Cespi (✉) · F. Passarini · E. Neri · F. Cavani
Department of Industrial Chemistry 'Toso Montanari', University of Bologna, Bologna, Italy
e-mail: daniele.cespi@gmail.com

R. Cucciniello
Department of Chemistry and Biology, University of Salerno, via Giovanni Paolo II, Salerno, Italy

S. Maranghi and C. Brondi (eds.), *Life Cycle Assessment in the Chemical Product Chain*, https://doi.org/10.1007/978-3-030-34424-5_3

3.1 Sustainable Green Chemical Industry

Creating and promoting a sustainable chemical industry means, at first, look at the concept of Green Chemistry. Green Chemistry represents the adoption of a set of principles, namely the twelve principles, able to reduce or eliminates the generation of hazardous substances during the entire life cycle of chemical products (design, manufacture, application and end-of-life). To date, the twelve principles (Anastas and Warner 1998) are internationally recognised and represent an essential reference for all the branches of chemistry (Fig. 3.1).

They are based on the fundamental idea of inspiring and guiding the scientific community and innovation towards environmentally friendly practices and solutions. During the last 20 years the field of Green Chemistry was spread out in many research fields pointing to a new and overwhelming direction for chemistry and a new and fascinating point of view for chemists (Anastas and Warner 1998). Chemists, as well as chemical engineers, are considered 'scientific designers'. Their role is to assemble new chemical structures without compromise the future of coming generations. They are called to develop a 'benign by design' chemistry (Anastas and Ferris 1994). In this sense, chemistry plays a pivotal role since has inspired the birth of *green chemists*. A new class of scientists animated by the necessity and the will to prevent environmental burdens within the chemical sector (e.g. waste, hazardous and non-renewable substances, pollution, energy depletion, non-degradability, etc.) (Erythropel et al. 2018).

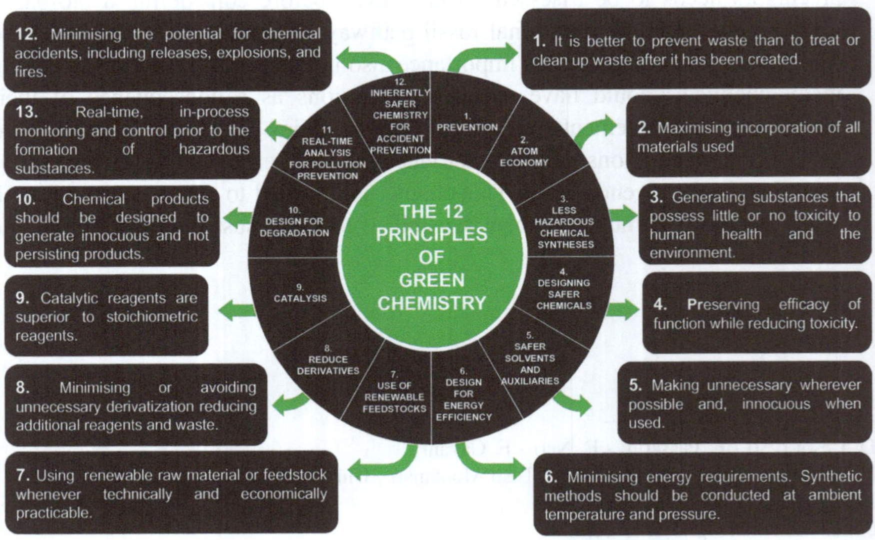

Fig. 3.1 The 12 principles of Green Chemistry

The sustainability behind the chemical industry is often questioned, since it contributes to the emission of impressive amount of greenhouse gases (GHGs)—184.1 Mt of CO_2 eq. in the US during 2017—and for the consumption of 15.6 Mt of TOE (tonnes of oil equivalent) in the EU (EPA 2017; Cefic 2018). Therefore, the development of sustainable chemistry has reached an increasing interest extended. The choice to pursue a sustainable approach needs and requires a holistic view which must ensure the protection and longevity of all living species (humans, animals and vegetables) taking into consideration economic aspects able to create business and profit, social and environmental impact of the value chain, global warming, problems of access to energy and so on (Marion et al. 2017).

3.2 Metrics in Green Chemistry

As suggested above Green Chemistry principles help to identify, qualitatively, the right way to build sustainable molecules. In the last years, researchers have developed simple metrics and indexes to assess the greener grade of reactions. At the beginning of 90s (Trost 1991) introduced the concept of *atom economy* (AE), which simply identifies how much of the reactant(s) ends up in the final product. This is possible by dividing molecular weight (MW) of the product for the cumulative MWs of all the reagents involved (sum). In the same years, Sheldon (1992, 1997) brought in the concept of *E-factor* (E_f) to point out the attention on the amount of waste generated per mass of product (in a step or reaction). Later in 2001 were introduced the *process mass intensity* (PMI) and *reaction mass efficiency* (RME). PMI is the ratio between the whole quantity of reagents used in a reaction and the amount of product generated (Curzons et al. 2001). In 2013, PMI was selected by the American Chemical Society Green Chemistry Institute Pharmaceutical Roundtable (ACS GCI PR) as key indicator to express sustainability in the pharmaceutical industry (Jiménez-González et al. 2011, 2013). RME expresses the ratio between the mass of target product with the sum of the masses for all reactants used. As demonstrated several times, all these metrics are strictly correlated each other and may be used single or combined to have an idea about the sustainability of a reaction (Andraos 2005; Lapkin and Constable 2009).

For example, GSK (GlaxoSmithKline plc) has applied the metrics above to compare impacts for different chemistries (Constable et al. 2002). They also included an economic analysis to understand the relationship between metrics and cost processes. Pfizer Inc. has published several studies (Dunn et al. 2004; Cespi et al. 2015a, b) in which AE, E_f and PMI were applied to a target molecule (sildenafil citrate) to show impacts reduction of its production over time and due to Green Chemistry improvements. PMI was also combined with LCA to show the trend of carbon footprint, embodied the energy and human health (Cespi et al. 2015a, b).

MANE (Phan et al. 2015) has released GREEN MOTION®, a tool developed by a free and peer-reviewed methodology able to evaluate the process greenness and critical issues grouping green metrics (e.g. E_f and PMI) and the twelve Green Chemistry principles to enable the measurement of the overall safety of a product, its impact on the environment and on the health of the people (Phan et al. 2015). In our recent work (Cespi et al. 2016a, b), we combined results from LCA with those achieved through GREEN MOTION® to assess the sustainability of an innovative bio-based route developed at a laboratory scale. In 2018, ACS GCI PR and IQ Consortium (https://iqconsortium.org/) launched the Innovation Green Aspiration Level (iGAL) methodology (Roschangar et al. 2018). iGAL is aimed to 'capture the impact of green process inventiveness and improvements, making it a useful innovation-driven green metric'.

However, the definition of what is inherently green is something that goes beyond the simple application of green metrics. Green metrics may help, but the procedure is rather than easy work. Very often, only using a life-cycle assessment, it is possible to identify the higher sustainability level of a process (Jiménez-Gonzàlez and Constable 2011; Anastas and Zimmerman 2003). The optimisation procedure is driven by the application of the fundamental principles, which look at the entire life cycle of a product (Gilbertson et al. 2015).

LCA methodology is often intended as a tool supporting one of the Green Chemistry principles (3: Less hazardous synthesis) (Erythropel et al. 2018), but also as a more comprehensive instrument of interpretation of the overall sustainability of new synthetic processes, and this is the topic widely addressed below. Indeed, as recalled by Erythropel et al. (2018) it could be restrictive to apply only one of the 12 principles, because an overall sustainability evaluation risks being lost: 'While the 12 Principles have been a useful framework, there needs to be a universal understanding that these are not twelve independent factors but rather an interconnected system by which design synergies can be imagined and realized'. As Sheldon confirms (Sheldon 2018), citing Graedel (1999), 'adding a life-cycle perspective to Green Chemistry enlarges its scope and enhances its environmental benefits'. For instance, Corona et al. (2018) show that Alfalfa (*Medicago sativa*), after undergoing a fractionation (shredding and pressing, aimed at separating the liquid stream i.e. the press-juice, from the solid fibre-rich fraction, named press-pulp), can be further used in different applications, which strongly affect environmental impacts related to the products that can be replaced, so that a greater product yield results in higher environmental benefits compared to a lower energy consumption. In this case, a 'conflict' between the principle 6 (Design for energy efficiency) and 1 (waste prevention) can be observed, in which the latter prevails; however, this conundrum could be solved only from a comprehensive perspective, as an LCA approach. The next paragraph deals with the description of some interesting applications of LCA in the industry to promote green and sustainable chemistry.

3.3 Life-Cycle Approach to Develop a Green Chemical Industry

Pioneers in Green Chemistry have recognised the importance of adopting a holistic approach since 2000s, suggesting the usage of LCA to identify greener routes/products (Aresta and Galatola 1999; Anastas and Lankey 2000; Doménech et al. 2002; Hellweg et al. 2004). However, one of the first attempts of a rough life-cycle analysis to the chemical sector is dated 1968 when Harold Smith presented a contribution entitled 'The cumulative energy requirements of some final products of the chemical industry' to the World Power Conference, in Moscow (Smith 1968). Smith has presented an elaborated assessment of what we call now *CED—Cumulative Energy Demand* of the major commodities of 60s, by including inorganic (phosphorus, sulphuric acid, soda ash, chlorine, ammonia, nitric acid and ammonium nitrate), organic (acetylene, vinyl chloride, methanol, benzene, xylenes, ethylene, styrene and butadiene) and polymeric (PVC, polyester fibre, PE and synthetic rubber) compounds. Before Smith, other researchers tried to expand the value of goods (not only chemical substances) behind the common *factory gate*. The Italian professor Roberto Salvadori highlighted the importance of considering the CED of goods to establish their real price (Salvadori 1930). From then on, the industry has started considering embodied environmental costs behind chemicals production. Nowadays, all the major corporates working in the field of chemistry (and not only) include life-cycle evaluations in their organisational structure. In this section, a list of the main LCA-based tools developed by companies to support their strategies is presented (Table 3.1). The list would not be considered as exhaustive since only based on the available literature, but it would represent an attempt to clarify the role of LCA inside the chemical industry.

Table 3.1 List of the leading tools developed by chemical companies to support their strategies

Developer company	Tools	Main function
L'Oréal	Eco-footprint tool	LCA outputs with green metrics
Solvay	Sustainable Portfolio Management (SPM) tool	Evaluation of environmental and economic impact
GlaxoSmithKline plc (GSK)	Eco-Design Toolkit ©	Support chemical designer in developing a benign by design chemistry
	FLASC™	Predict potential cradle-to-gate impacts of new chemical routes
Bayer AG	Interactive LCA Calculator Tool	Evaluate the differences in environmental impact comparing different alternatives
National Risk Management Research Laboratory—Office R&D	GREENSCOPE	Support in developing more sustainable chemical reaction

First example here reported is the *Eco-footprint tool*, a combination of simple LCA outputs with green metrics developed by Chimex, a subsidiary of L'Oréal (Leseurre et al. 2014). Since 2012 the L'Oréal Research & Innovation Center has pointed out its commitment to developing a green and sustainable chemical industry by combining eco-design and Green Chemistry principles with simple green metrics (Philippe et al. 2012). In 2014, they launched the Eco-footprint tool, a gate-to-gate approach which provides a valuable instrument to the syntheses optimisation and the communication of the environmental performance. The tool estimates both the eco-design and the manufacturing footprints, respectively, based on R&D and production plants data. Results are expressed in terms of 10 different indicators. Among these, the carbon footprint is also included to quantify burdens due to energetic consumption in the production plant. Tool converts data concerning energy use (electricity, steam, fuel, cooling, liquids, etc.,) and carbon emissions into CO_2 eq. automatically, through emission factors. The other nine indicators are water consumption, raw materials geographical origin, aqueous waste valorisation, used organic solvents valorisation, synthetic pathway efficiency, the raw material of renewable origin, Ef, the potential environmental impact of raw materials and potential environmental impact of waste.

On that subject, at Beiersdorf, LCA is used to drive R&D in the development of safer and more environmentally friendly products. Accordingly, to their data, the major environmental concerns resulting from their products are outside the company boundaries and they are concentrated on the acquisition of the raw materials (basic chemicals and packaging) (Beiersdorf AG 2019). Therefore, the life-cycle perspective is aimed to rethink their supply chain, moving from one to another supplier which demonstrates a greater environmental awareness by the possession of ISO labels and schemes (e.g. 14001, 14025, etc.,) or by the use of recycled materials. LCA is also necessary to switch from one material to another. This is the example of NIVEA, a Beiersdorf's brand for body care. In such case, the company has found that using plastic jars instead of glass could reduce the potential impact of greenhouse gases (GHGs) around 16–28% (depending on PET or PP) (Beiersdorf AG 2019). Final consumers, of course, have the lion's share of credit since results strongly depend on their behaviour: If they do not recycle empty plastic jars this may have negative effects on other environmental aspects (damage of ecosystems and fossil resources depletion) (Swarr et al. 2019).

Switching from personal care to polymers, the usage of LCA to support R&D is also carried out at Covestro AG. Covestro applies iterative LCAs to analyse hotspots occurring during early design activities. This strategy is aimed to promote continuous improvement before market lunch. LCA analyses could be simple *screening* or *full* depending on the severity, and they might lead to project cancellation if it is not sustainable enough (Benetto et al. 2018). In Covestro, LCA is also used as a benchmark to counterpose environmental advantages over competitors (Covestro 2019).

Another example of LCA used in industry to treat hotspots and support R&D is the P&G (The Procter & Gamble Company) strategy to assess the *Company Footprint*. Through the analysis of some of the major indicators (e.g. CO_2 eq., energy and water demand/use, eutrophication, etc.) P&G environmental managers can identify the life-cycle stages with deeper environmental concerns. Based on those results they planned to implement supply chain and/or product design. An example is the development of cold washing products given that results cradle-to-grave LCA analyses have shown great impacts due to the energy consumption during product usage (e.g. laundry/dishwasher) (P&G 2019; Benedetto et al. 2018).

Solvay has launched the Sustainable Portfolio Management (SPM) tool, a global and systematic assessment aimed to anticipate the impact of corporate business by covering the entire product's life cycle (Solvay S.A. 2019). SPM tool combines environmental and economic impacts. First, upstream stages are analysed through cradle-to-gate LCA by considering 19 impact indicators. Results are then monetised with *shadow costs*, which represent the hypothetical costs to remedy or prevent negative impacts. Scores achieved are compared to sales value to obtain Operations Vulnerability. The downstream (gate-to-grave) is based on market signals on sustainability trends collected from regulators, NGOs, IOs, clients, consumers, etc. At the end of the entire procedure SPM tool is able to classify products as Solutions (products with better sustainability contribution to Solvay customers and value chain, combined with a favourable balance between value and environmental impact), Neutral (products that consumers need, but which do not contribute to environmental footprint reductions) and Challenges (products with significant negative impact on revenue over time and where products may eventually disappear).

The application of the holist approach to support decision-making and to drive the company's strategy is also done by Henkel AG & Co KGaA. At Henkel life-cycle implication of products, services and operation are evaluated and, where robust data are available impacts are translated into externalities. The carbon footprint represents the main indicator selected to communicate company environmental performances in their annual Sustainability Report (available for stakeholders). The assessment is extended to all the three *Scopes* (EN ISO 14064:2012) by including direct and indirect emissions along the entire life cycle. In the 2017 report they disclosed that around 98% of the total operational carbon footprint is due to indirect emission: the majority (68%) is due to usage stage (consumers' behaviour); production and distribution only affect the 29% of the whole impact (Henkel AG & Co KGaA 2019). Results from carbon footprint are then translated into externalities since helpful for the company to evaluate risks and opportunities and thus allocate financial resources to promote sustainable development (Henkel AG & Co KGaA 2019). In general, they used different sources (World Bank Group 2016) to estimate the cost of CO_2 eq. since carbon prices may vary significantly. Quantification of the externalities is a hard and challenging procedure, full of uncertainties. However, the use of different sources combined with their methodology sounds like a scientific and reproducible approach. In addition to CO_2 eq., Henkel is also studying methodologies to include other indicators in the evaluation of sustainability. The water footprint is right now evaluated by the use of the Water Scarcity method (Berger et al. 2014). A material footprint index would be also be investigated through the development of a new indicator able to consider both the use of finite fossil and mineral resources as well as renewable resources and their potential impact on biodiversity. This would be the result of a combination of the Agricultural Land use Occupation Potential and Abiotic Depletion Potential Henkel AG & Co KGaA (2019).

Pharmaceuticals companies represent another branch of the chemical industry that considers such approaches in their core business. In 1998, GlaxoSmithKline plc (GSK) had launched the Eco-Design Toolkit© (Curzons et al. 1999) to support 'chemical designer' in developing a benign by design chemistry (Anastas and Ferris 1994) in the manufacture of active pharmaceutical ingredients (API) (Henderson et al. 2011). Later in 2004, they published their first example of LCA application into production activities and in 2007 they released the FLASC™ (acronyms of Fast Life Cycle Assessment of Synthetic Chemistry) (Jiménez-González et al. 2004; Curzons et al. 2007; Adams et al. 2013). This web-based tool is a simple methodology able to support R&D activities, by promoting fast and simple (not simplistic) evaluation at an early stage. The application of FLASC™ is promoted by GSK to support business development choices and help researchers to predict potential cradle-to-gate impacts of new routes under development. Benchmarking between existing or proposed routes or processes is allowed. Results trend is shown from medicinal to optimised/commercial routes by supporting EHS experts on the choice of materials with greatest life-cycle performances.

At Merck KGaA the life-cycle thinking is used to support key decisions and actions (Merck KGaA 2019). LCA guarantees transparency in understanding which areas represent the most impactful. Results from LCA help R&D researchers to design and study new solutions which allow consumers to reduce their burdens. This is the case of Lab Water Elix® Water Purification System (Merck KGaA 2019): LCA has alerted researchers they have to work to guarantee a reduced water consumption during usage. LCA is an intense process, therefore they have developed a series of guidelines to ensure consistency and help practitioners in the choice of data quality, system boundaries, how to treat recycled materials, etc. In general, Merck used carbon footprint to communicate the environmental sustainability of their products. However, it depends on the project objectives. For example, in the case of MilliFlex® Quantum product (Merck KGaA 2019) LCA was used to support the design of new packaging solutions: by re-thinking shape and material layers researchers have reduced impacts during shipping and conservation. Screening-level LCAs are adopted for benchmarking: New ideas are counterposed to traditional products to drive innovation in the right way (e.g. Clarisolve® depth filters vs. Millistak+® POD depth filters) (Merck KGaA 2019).

Life-cycle-approach is a part of Bayer AG core business. The application of LCA in Bayer is aimed to support the research of new and cleaner chemicals/routes, but it is not limited to R&D only. In 2011 it was developed the *Interactive LCA Calculator Tool* (Bayer AG 2019) (property of Bayer; designed and powered by Quantis International). This web-based LCA tool was thought to show differences in environmental impacts of Long-Lasting Insecticidal Nets (LLINs) to fight malaria in developing countries. The tool compares three alternative types of

LLINs: a polyethylene net, a polyester net and a polypropylene net (LifeNet® the product property of Bayer). The analysis may consider two different scenarios (20 or 30 washes of the net), the country and the number of installations. Results are obtained by the IMPACT 2002+ model (Jolliet et al. 2003) which expresses them into four different indicators (global warming, water withdrawal, resource consumption and ecosystem quality). The social purpose behind this LCA utilisation is clear when further considerations are carried out. Developing countries need efficient systems, which require a lower amount of resources with minimal production of waste to prevent further pollution. Also, results from the LCA analysis were also used by the company to obtain the EPD (Environmental Product Declaration) certificate for the LifeNet®. EPD is a Type III eco-label for products and services, normalised by the ISO 14025:2010 (EN ISO 14025:2010). It is based on LCA analysis and represents one of the most consolidated international certificates with 3–5 years validity worldwide. Its importance to promote a more sustainable chemical industry was the object of a recent publication (Swarr et al. 2019). Not only Bayer has used results from LCA analysis to obtain EPDs of products. Other examples of great chemical corporates are Akzo Nobel N.V., PPG Industries, Inc. and Sherwin-Williams Company (the list is not intended to be exhaustive).

In conclusion, in this paragraph, we tried to present a brief list of industries which are currently applying a life-cycle perspective in their day-by-day activities. However, the application of LCA-based tools is not only limited to private companies but extended to universities and governmental agencies as well. An example is a work done by EPA (US Environmental Protection Agency) to support the development of greener processes. In 2003, the National Risk Management Research Laboratory—Office R&D released an indicator model called GREENSCOPE (Gauging Reaction Effectiveness for Environmental Sustainability of Chemistries with a Multi-objective Process Evaluator) (Gonzalez and Smith 2003). This tool laid the basis for the development of more sustainable chemical reaction or process by investigating four areas of interest (environment, energy, efficiency and economics) by combining eight major indicators into a unique analysis methodology. Some examples of GREENSCOPE application were already published in the peer-review journal (Ruiz-Mercado et al. 2012a, b, 2016).

3.4 Bio-based Industry and Life-Cycle Approach

A sector of tremendous resonance for the entire sector of chemistry is represented by the substitution of fossil raw materials with bio-based building blocks. This concept, argued in the 7th principle of Green Chemistry, is expected to grow more in the next years. The use of bio-based feedstocks (in the so-called bio-based industry) is predicted to surpass that of products obtained by non-renewable fossil sources (Clark et al. 2015). Bio-based raw materials represent a consistent source of renewable carbon; limitations of geological resources and CO_2 emissions and their impact on the global warming and climate changes are driving a revolutionary transition towards bio-based industry (Ricciardi et al. 2017a, b, 2018; Tripodi et al. 2017). The list of bio-based chemicals has been recently updated and some of these chemicals are commercialised in large scale. Some examples include itaconic acid, succinic acid, acetic acid, 1,2- and 1,3-propanediol, 1,2-pentandiol, sorbitol, furfural, xylitol, ethylene glycol and epichlorohydrin (Marion et al. 2017).

The continuous development of biorefineries is naturally concomitant with the depletion of oil driving a significant reorganisation of the entire chemical industry. As a matter of fact, oil refineries are in the process of being redesigned coupling petrochemical processes with bio-based productions and fermentation technologies. Some relevant examples are briefly introduced to highlight an exponential growth during the last twenty years and to presage a forthcoming pivotal role in the chemical industry.

The most consolidated and well-known use of biomass is represented by the production of fuels, an alternative to oil for feeding the transportation sector.

In this scenario, Bio-Synfining (property of Syntroleum Corporation) has developed industrial processes currently available which can transform vegetable or animal oils, fats and greases into renewable synthetic fuels such as diesel, naphtha and propane (Abhari et al. 2013).

Other companies producing biofuels from biomass are Total Petrochemicals Inc., Eni S.p.A.-Honeywell UOP (Ecofining), Solena fuels (GreenSky), Rentech, the Energy and Environmental Research Center (EERC), Finland's UPM and the Renewable Energy Group (REG). Furthermore, Sasol Limited and the Royal Dutch Shell plc have developed biomethane routes to obtain bio-naphtha (Vermeiren and Van Gyseghem 2011).

Switching from fuels to building blocks, an important example for polymers production (mainly polyethylene) is represented by ethylene with a production of 150 Mt in 2016. Solvay, BP plc and Total Petrochemicals Inc. have patented several processes to obtain bio-ethylene through bio-ethanol dehydration (Minoux et al. 2009; Fukumoto and Kimura 2013; Patrick Gracey and Partington 2009).

Dow Global Technologies has patented a new promising process to produce propylene using glycerol, mainly obtained as a by-product from bio-diesel preparation, as starting material in the presence of HI (hydroiodic acid) and hydrogen in a batch reactor (Deshpande et al. 2013). Also, the production of diols (a chemical compound containing two hydroxyl groups, HO-R-OH) from biomass seems to be important for some chemical corporates, since these start molecules can be used in the bio-olefins supply chain. LanzaTech NZ Ltd., Versalis S.p.A.-Genomatica Inc., Genecor International, Inc. (a subsidiary of DuPont)—The Goodyear Tire & Rubber Company, and Global Bioenergies SA are developing several biological synthesis routes to produce C4 diols (i.e. butanediols) from carbohydrates (Lanzafame et al. 2014). C4 diols and alcohols (for example, *n*-butanol, which is obtained from ethanol through the Guerbet reaction can be used in turn to prepare bio-olefins (butadienes and 1-butene) (Aitchison et al. 2016).

In 2018 Danone SA, PepsiCo Inc., Nestlé Waters (the water division of the Nestlé Group SA) and Origin Materials have established an R&D partnership to develop 100% bio-based bottles made from bio-based polyethylene terephthalate

(PET) (Lichtlen and Pochard 2018). This is possible through the coupling of bio-based mono ethylene glycol (MEG) and terephthalic acid obtained in turn from bio-based *p*-xylene. As a matter of fact, Virent Inc. is producing bio-*p*-xylene, branded BioFormPX, through carbohydrates conversion, whereas Gevo Inc. has developed an industrial process to obtain bio-*p*-xylene by isobutanol conversion (Gevo® route) (Cortright and Blommel 2013).

Another important bio-based building block is 2,5-furandicarboxylic acid (FDCA), synthesised by the catalytic oxidation of 5-hydroxymethylfurfural (HMF), esters and ethers of HMF. It is produced by Avantium NV with a year capacity between 30,000 and 50,000 t. FDCA-derived polyethylene furanoate (PEF) has comparable mechanical and barrier properties several times higher than those of PET and is expected to be used for soft drink and beer bottle production; Coca-Cola, Ford Moto, Heinz, Nike and P&G have planned to realise this strategy (Avantium 2019).

Another example of bio-based packaging is represented by the work done at Novamont S.p.A., an Italian chemical company that produces a biodegradable polymer namely MaterBi®. It is obtained from starch; the latest polymer type includes azelaic acid and 1,4-butanediol as bio-based monomers (Novamont 2019).

The production of bio monomers for fibres is important as well. Rennovia Inc. has developed a bio-based route to prepare adipic acid used in the preparation of nylon-6,6. The process uses two catalytic steps that involve the aerobic oxidation of glucose to glucaric acid followed by hydrodeoxygenation of glucaric acid to adipic acid (Rennovia 2019).

It also seems possible to convert solid biomass directly into aromatic products. The latter offers several advantages because lignin is the second most abundant fraction of the lignocellulose and for the high value of the obtained products. This strategy developed by Anellotech Inc. (an innovative spin-off based in the US) is based on the use of fast catalytic pyrolysis (CFP) of the solid parts of biomass into aromatic compounds as major products (Huber et al. 2012). More in detail, during the CFP process, the biomass decomposes thermally into vapours, and the cellulose and hemicellulose form anhydrosugars via pyrolysis.

Another example also represented by the Epicerol® process developed by Solvay to prepare bio-epichlorohydrin (ECH) from glycerol. ECH is an essential feedstock for the production of epoxy resins (used in coatings, adhesives, composite and electrical insulation materials). The Epicerol® process represents a successful example of bio-based industrial plant, an alternative to the propylene-based ECH production (Cespi et al. 2016a, b).

Notwithstanding all this, the use of renewables as starting materials for the chemical industry currently is not sufficient, and it does not guarantee sustainable chemistry. For example, the use of biomass, a renewable material (in line with principle 7), is not always more advantageous in environmental terms, compared to fossil raw materials. This usually may occur if the biomass comes from dedicated crops and not from by-products; in this case, there is a dilemma between the principles 2 (Atom Economy) and 7, since the higher yield of a conventional petrochemical process could globally result in a lower environmental burden, in comparison with an alternative route from biomass, particularly if the latter comes from impacting agricultural activities.

Furthermore, depending on the biomass type (sugar cane, corn, wheat, rye, sugar beet) the same bio-based product (ethanol) could result more or less impacting and more or less suitable compared to the fossil-based product. Also, in this case, a conflict between the principles 7 and 1 (but also 2) may occur.

This means that chemical industries oriented to bio-based solutions, interested in following a life-cycle perspective to support their R&D and business development, would consider the approach here suggested and proposed by the more recent scientific literature.

Van Schoubroeck et al., describe the main sustainability indicators used for bio-based chemical compounds. All indicators are considered, included those describing economic and social sustainability. Authors state that in the examined literature, papers dealing with a set of indicators at only 1 dimension, are all relating to environmental sustainability (50%), while those at 2 dimensions also include economic considerations (34%), thus only 16% includes social ones, i.e. they consider all the three dimensions (Van Schoubroeck et al. 2018).

These are major environmental indicators: Climate mitigation, Clean and efficient energy use, Resource Management, Ecosystem care. The most used is GWP, expressed in CO_2 equivalents: all climate-changing substances emitted are added, subtracting the CO_2 fixed by the biomass during its growth. The second largely employed environmental indicator is the Cumulative Energy Demand (CED), defined as 'the total direct energy use throughout the entire life cycle'. Another indicator concerns resources management, which could result in a two-fold outcome: on the one hand the use of renewable material from biomass appears favourable (mostly for the production of chemicals) compared to the fossil one; on the other, frequently a critical parameter occurs, consisting in 'land use' (i.e. occupation) o 'land use change' (or transformation). It is recalled that 'the ReCiPe method includes urban and agricultural land occupation as well as natural land transformation to calculate the full land use impact'. More seldom, the Indirect Land Use Change (ILUC) indicator is introduced, relating to the climate-changing emissions due to the variation in soil use. Finally, the fourth set of indicators (generally called 'ecosystem care') takes into account the impacts due to soil degradation and pollutants introduction: air pollution (including photo-oxidant formation), (marine and freshwater) eutrophication, acidification, ecotoxicity and waste generation.

Different indicators are also reported in Patel et al., in a multidimensional view (Patel et al. 2012). Those related to environment are divided in 'Environmental impact (EI) of raw materials' (CED and GHG emissions), 'Process costs and environmental impact (PCEI)' (mainly connected to process material and energy flows) and 'Environment Health Safety index' (including output flows of chemicals, thus persistency, air hazard, water hazard, solid waste parameters). However, the global value comes from the integration of all indicators, with a multi-criteria approach of the weighting of the different variables.

Weiss et al. consider 6 environmental impact categories of different molecules (also polymeric) synthesised from bio-based materials and compared to conventional ones: non-renewable energy use, climate change, eutrophication, acidification, stratospheric ozone depletion, photochemical ozone formation (Weiss et al. 2012). It is observed that other studies consider less common impact categories, namely human and terrestrial ecotoxicity and carcinogenic potentials, in which conventional materials are usually more impacting, while in the case of aquatic ecotoxicity bio-based materials typically display higher impacts. Ultimately, the comparison between bio-based and conventional materials (considering a scope 'from cradle to gate') leads to different results, depending on impact categories, materials under investigation, type of cultivation (intensive, extensive) and other variables more associated with territorial peculiarities. However, in general, bio-based compounds provide greater benefits particularly in the categories GWP, CED and photochemical ozone formation, while result in higher impacts in the categories eutrophication and stratospheric ozone depletion (the latter associated to N_2O emissions); more debated is the best option concerning the category acidification.

Finally, an interesting application of LCA to a cosmetic product realised with bio-based ingredients (Secchi et al. 2016), shows that the use of natural products not necessarily results in an improvement of the overall environmental performance, because it depends on the processes that input materials must undergo (especially liquid fraction saponification to eliminate the unsaturated oleic/linoleic triglycerides fraction). As the case may be, bio-based scenarios could show higher impacts in the categories GWP, human toxicity, photochemical ozone formation, acidification, freshwater eutrophication and ecotoxicity and water resource depletion. However, this confirms, as said above, that sometimes two Green Chemistry principles could conflict and one prevails, considering the benefits from a global perspective: in this case, principle 5 (Safer solvents and auxiliaries), particularly for the use of KOH, prevails on principle 7 (Use of renewable feedstock).

3.5 Social-LCA in the Chemical Industry

The social life-cycle assessment (S-LCA) is defined as 'a social impact (and potential impact) assessment technique that aims to assess the social and socio-economic aspects of products and their potential positive and negative impacts along their life cycle encompassing extraction and processing of raw materials, manufacturing, distribution, use, re-use, maintenance, recycling and final disposal' (UNEP/SETAC 2009).

In 2009 has been published by UNEP SETAC Life Cycle Initiative (LCI) the guidelines that represent a milestone for the social life-cycle assessment literature (UNEP/SETAC 2009).

Then, in 2013, for the first time, a group of seven companies decided to develop a 'team work', with the support of PRé Consultants (PRé Consultants 2019), to develop and promote the assessment of the social impact of the life cycle of a product. In 2014, from the collaboration has been published the first draft of the Handbook by the Roundtable for Product Social Metrics (Traverso et al. 2018), update with a new version in 2016 (Fontes et al. 2014).

The main key point of the social assessment methodology is that the benefit target taken into account in an S-LCA may be divided into stakeholders groups: workers, consumers, local community, society and value chain actors and for every stakeholder group are taken into account some of the most relevant topics regarding the category (UNEP 2013). For example, for the 'consumers' stakeholder category, the main aspects taken into account are health and safety, feedback mechanism, privacy, transparency and end-of-life responsibility. In an S-LCA impact categories may be assessed using different indicators depending on the stakeholder group under consideration, and the subcategories are defined as 'socially significant themes according to the stakeholder's group' (Traverso et al. 2018).

In contrast to environmental LCA, where environmental impacts are quantitatively linked to a functional unit, so in S-LCA social indicators are used to characterise social implications in which the activity described by the functional unit occurs (Benoit-Norris 2012).

Companies should manage with the social assessment of their activity (i) to address to strategic issues and issues posed by the marketplace, politics and research; (ii) to minimise the business risk of losing reputation and their licence to operate.

The relevance to also assessing the social impact in every production area, therefore even along the ideation and scale-up of a chemical process is highlighted at the international level. In 2015, UNO (United Nation Organisation) described in the Agenda 3030 for a Sustainable Development, adopted by all United Nations Member States, the 17 Sustainable Development Goals (SDGs 2019) to pursue peace and prosperity for people and the planet, now and into the future (SDGs 2019). The goals represent an urgent call for action by all the countries in a global partnership, with strategies to improve health, education, reduce inequality, with a strong goal through tackling climate change and preserve the ecosystems of the planet.

Most of them include the evaluation of the social impacts, with the goals to achieve in the future no poverty, zero hunger, good health and well-being, good quality for the education, the reduction of the inequalities with peace and justice.

The principles described above to pursuit the sustainable development should be applied in every design of the new system and process, so even in the design of new chemical products. Due to that, more and more several companies are applying, in addition to the Green Chemistry principles, also the study of the theory and methodology to assess the social impact related to their production.

Among the main chemical companies, a great contribution to the development and usage of social implications linked to their business activity is attributed to BASF SE. BASF has declared its support for the global initiatives of Sustainable

Development and Responsible Care and is a member of the World Business Council of Sustainable Development and the Global Compact initiative of the United Nations (Shonnard et al. 2003).

In 1995, researchers at BASF have developed a strong methodology for the application of Social-LCA and Social Hotspot Analysis (including an SDG Evaluation to Product Assessment) so-called SEEbalance® (Schmidt et al. 2004). Moreover, SEEbalance® is a comparative life-cycle assessment tool that consists of three main aspects: costs, environmental impact and social effects of different product or process alternatives examined and compared.

BASF, together with Karlsruhe University and the Öko-Institut e.V. (both in Germany), has developed this tool for the measurement of social aspects to incorporate them into its existing eco-efficiency analysis that considers the economic and environmental effects of a product giving these equal weighting.

The overall interpretation of the results is the combination of the Eco-Efficiency Analysis and the Social Analysis because the methodology assesses the ecological and economic impact of alternative products and processes. Regarding the Social Analysis, the methodology considers as impact categories those defined in the outcome of the 'Round table for Product Social Metrics', assessing 11 impacts categories for three different groups of stakeholders: workers, local communities and consumers.

At the beginning of an eco-efficiency analysis, there is some basic precondition to take into account (Saling et al. 2002, 2005). The main theme is that the final customer benefit is at the heart of the analysis and all products or processes studied have to meet the same customer benefit considering the entire life cycle and assessing both ecological and economic issues in addition to the impact on health and the danger to people. After that, in the Hotspot Assessment, the SDGs (2019) are used as an orientation for the detection of Social Hotspots in a value chain.

The results obtained from the methodology may be used by accompanying as BASF mainly to support strategic management and compare strengths and weaknesses about competitors, due to the optimisation of products and processes through the design of new eco-efficient products to be launched on the market.

The example of BASF is not the only one in the chemical sector.

At the research level, a virtuous example is the application of social impact assessment to nano-enabled products (Subramaniam et al. 2018). In the study the S-LCA and the multi-criteria decision analysis (MCDA) are integrated with four steps of the analysis: normalisation of company level data on the social indicator, weighting at stakeholder and indicator level, aggregation of social indicators score into benefit score and cost classification. The results obtained could be then linked to the outputs of the environmental and economic assessment to have an overall sustainability assessment.

In the industry, another integrated approach of social and environmental assessment has been promoted, for example, by Solvay through the Sustainable Guar Initiative, launched in 2015 (SGI). Sustainable Guar Initiative (SGI) is a three-year-long integrated programme aiming at developing sustainable guar production within the Bikaner district Rajasthan, in India. Guar gum is extracted from guar seeds and can be used as such, or functionalised. It can be used as a bio-based thickening/conditioning agent in personal care applications. SGI was set up by Solvay, l'Oreal, HiCherm and the NGO TechnoServe, and is based on four pillars:

- agronomy: enhancing sustainable practices for rain-fed guar production;
- environment: groundwater-neutral approaches and best practices in guar farming, along with tree plantations;
- social impact: gender approaches, nutrition, health and hygiene;
- market improvement: traceability, supply chain and market access.

SGI was used as a case study to structure social impact assessment methodology on smallholders and can be considered as a guiding document for future application of social life-cycle assessment.

Conclusion and Remarks

As reported in this chapter, the application of LCA-based methodologies is well diffused in chemical industries, in particular in companies which promote Green Chemistry. The life-cycle perspective is useful to support R&D and business development strategies aimed to achieve the common target of sustainability.

LCA could be used alone or in combination with other tools. It has also been shown that LCA and Green Chemistry Metrics could fruitfully be employed together 'to evaluate and guide fine chemical synthesis' (Lee et al. 2018) and provide complementary and useful indications to industries and researchers.

Sometimes it could be particularly difficult and time-consuming to carry out a complete LCA study from the cradle to the grave in the stage of design and development of new substances; depending on the target, the analysis could be limited to the phases from cradle-to-gate.

In addition to that, some points remain tricky in the integration of LCA methodology with Green Chemistry principles (Muñoz 2012). For instance, the qualitative nature of Green Chemistry principles is sometimes seen as critical. However, there are barriers to deeper penetration of LCA, as it is perceived as time-consuming and costly, many inventory data for chemical compounds are still lacking, primary data are missing in the starting developing phase, its focus is on potential instead of effective impacts and risks.

The problem remains of quantifying the benefits of a new reaction, switching from the laboratory to an industrial scale, especially for the newest applications, as nanotechnologies, where databases lack more (Kralisch et al. 2015). In this regard, Chaps. 5 and 6 attempts to describe the main approaches that can be useful to overcome these drawbacks. Furthermore, the ever-increasing publications on this topic provide more and more data available and thus to improve also predictive tools, able to help who is designing new compounds by a prior assessment (Erythropel et al. 2018).

In addition, some Green Chemistry principles likely are beyond LCA considerations, like 11 or 12 (Real-time analysis for pollution prevention; Inherently safer chemistry for accident prevention).

Nevertheless, it is widely recognised that the application of a life-cycle thinking perspective may represent the right recipe to address sustainability in the chemical sector.

References

Abhari R, Tomlinson HL, Roth G (2013) Biorenewable naphtha. US Patent 8558042B2, Syntroleum

Adams JP, Alder CM, Andrews I et al (2013) Development of GSK's reagent guides—embedding sustainability into reagent selection. Green Chem 15:1542–1549

Aitchison H, Wingad RL, Fass DL (2016) Homogeneous ethanol to butanol catalysis guerbet renewed. ACS Catal 6:7125–7132

Anastas PT, Ferris CA (1994) Benign by design. Alternative synthetic design for pollution prevention, vol 577. American Chemical Society

Anastas PT, Lankey RL (2000) Life cycle assessment and green chemistry: the yin and yang of industrial ecology. Green Chem 2:289–295

Anastas PT, Warner JC (1998) Green chemistry: theory and practice. Oxford University Press, USA

Anastas PT, Zimmerman JB (2003) Peer reviewed: design through the 12 principles of green engineering. Environ Sci Techol 37:94–101

Andraos J (2005) Unification of reaction metrics for green chemistry: applications to reaction analysis. Org Process Res Dev 9:149–163

Aresta M, Galatola M (1999) Life cycle analysis applied to the assessment of the environmental impact of alternative synthetic processes. The dimethylcarbonate case: part 1. J Clean Prod 7:181–193

Avantium (2019). https://www.avantium.com/yxy/products-applications/. Accessed 28 Feb 2019

Bayer AG (2019). http://lifenet.bayer.fr/. Accessed 28 Feb 2019

Beiersdorf AG (2019). https://www.beiersdorf.com/sustainability/products/packaging/life-cycle-analyses. Accessed 28 Feb 2019

Benetto E, Gericke K, Guiton M (2018), Designing sustainable technologies, products and policies. https://doi.org/10.1007/978-3-319-66981-6_39. Accessed 28 Feb 2019

Benoit-Norris C (2012) Social life cycle assessment: a technique providing a new wealth of information to inform sustainability-related decision making. In: Curran MA (ed) Life cycle assessment handbook: a guide for environmentally sustainable products. Wiley, Hoboken, p 433

Berger M, van der Ent R, Eisner S, Bach V, Finkbeiner M (2014) Water Accounting and Vulnerability Evaluation (WAVE): considering atmospheric evaporation recycling and the risk of freshwater depletion in water footprinting. Environ Sci Technol 48:4521–4528

Cefic (2018). http://www.cefic.org/Documents/RESOURCES/Reports-and-Brochu-re/Cefic_FactsAnd_Figures_2018_Industrial_BROCHURE_TRADE.pdf. Accessed 28 Feb 2019

Cespi D, Passarini F, Mastragostino G et al (2015a) Glycerol as feedstock in the synthesis of chemicals: a life cycle analysis for acrolein production. Green Chem 17:343–355

Cespi D, Beach ES, Swarr TE et al (2015b) Life cycle inventory improvement in the pharmaceutical sector: assessment of the sustainability combining PMI and LCA tools. Green Chem 17:3390–3400

Cespi D, Cucciniello R, Ricciardi M et al (2016a) A simplified early stage assessment of process intensification: glycidol as a value-added product from epichlorohydrin industry wastes. Green Chem 18:4559–4570

Cespi D, Passarini F, Vassura I et al (2016b) Butadiene from biomass, a life cycle perspective to address sustainability in the chemical industry. Green Chem 18:1625–1638

Clark JH, Farmer TJ, Hunt AJ et al (2015) Opportunities for bio-based solvents created as petrochemical and fuel products transition towards renewable resources. Int J Mol Sci 16:17101–17159

Constable DJC, Curzons AD, Cunningham VL (2002) Metrics to 'green' chemistry—which are the best? Green Chem 4:521–527

Corona A, Ambye-Jensen M, Vega GC et al (2018) Techno-environmental assessment of the green biorefinery concept: combining process simulation and life cycle assessment at an early design stage. Sci Total Environ 635:100–111

Cortright RC, Blommel PG (2013) Synthesis of liquid fuels and chemicals from oxygenated hydrocarbons. US Patent 2013/0185992 A1, assigned to Virent Inc.

Covestro AG (2019). https://www.productsafetyfirst.covestro.com/en/assessments/life-cycle-assessment. Accessed 28 Feb 2019

Curzons AD, Constable DJC, Mortimer DN et al (1999) Solvent selection guide: a guide to the integration of environmental, health and safety criteria into the selection of solvents. Clean Prod Process 1:82–90

Curzons AD, Constable DJC, Mortimer DN et al (2001) So you think your process is green, how do you know? Using principles of sustainability to determine what is green—a corporate perspective. Green Chem 3:1–6

Curzons AD, Jiménez-González C, Duncan AL et al (2007) Fast life cycle assessment of synthetic chemistry (FLASC™) tool. Int J Life Cycle Assess 12:272–280

Deshpande R, Davis P, Pandey V, Kore N (2013) Dehydroxylation of crude alcohol streams using a halogen-based catalyst. WO Patent 2013090076A1, Dow Global

Doménech X, Ayllón JA, Peral J (2002) How green is a chemical reaction? Application of LCA to Green Chemistry. Environ Sci Technol 36:5517–5520

Dunn PJ, Galvin S, Hettenbach K (2004) The development of an environmentally benign synthesis of sildenafil citrate (Viagra™) and its assessment by Green Chemistry metrics. Green Chem 6:43–48

EN ISO 14025 (2010) Environmental labels and declarations—type III environmental declarations—principles and procedures. International Organization for Standardization, Geneva, Switzerland

EN ISO 14064 (2012) Greenhouse gases—part 1: specification with guidance at the organization level for quantification and reporting of greenhouse gas emissions and removals. International Organization for Standardization, Geneva, Switzerland

Environmental Protection Agency (EPA) (2017). https://www.epa.gov/ghgreporting/ghgrp-chemicals

Erythropel H, Zimmerman JB, de Winter TM et al (2018) The Green ChemisTREE: 20 years after taking root with the 12 principles. Green Chem 20:1929–1961

Fontes J, Bolhuis A, Bogaers K et al (2014) Handbook of product social impact assessment. https://product-social-impact-assessment.com/. Accessed 28 Feb 2019

Fukumoto M, Kimura A (2013) Process for the manufacture of ethylene by dehydration of ethanol. EP Patent 2594546A1. Solvay S.A.

Gilbertson LM, Zimmerman JB, Plata DL (2015) Life cycle inventory improvement in the pharmaceutical sector: assessment of the sustainability combining PMI and LCA tools. Designing nanomaterials to maximize performance and minimize undesirable implications guided by the principles of Green Chemistry. Chem Soc Rev 44:5758–5777

Gonzalez MA, Smith RL (2003) A methodology to evaluate process sustainability. Environ Prog 22:269–276

Graedel TE (1999) Green chemistry in an industrial ecology context. Green Chem 1:G126–G128

Hellweg S, Fischer U, Scheringer M et al (2004) Environmental assessment of chemicals: methods and application to a case study of organic solvents. Green Chem 6:418–427

Henderson RK, Jiménez-González C, Constable DJC et al (2011) Expanding GSK's solvent selection guide—embedding sustainability into solvent selection starting at medicinal chemistry. Green Chem 13:854–862

Henkel AG & Co KGaA (2019). https://www.henkel.com/sustainability/positions/impact-measurement-and-valuation#Tab-805422_4. Accessed 28 Feb 2019

Huber GW, Gaffney AM, Jae J, Cheng YT (2012) Systems and processes for catalytic pyrolysis of biomass and hydrocarbonaceous materials for production of aromatics with optional olefin recycle, and catalysts having selected particle size for catalytic pyrolysis. US Patent 2012/0203042 A1, assigned to Anellotech Inc., and University of Massachusetts

IQ Consortium (2019). https://iqconsortium.org/. Accessed 28 Feb 2019

Jiménez-González C, Constable DJC (2011) Green Chemistry and engineering: a practical design approach. Wiley-VCH Verlag Gmbh & CO. KGAA, Weinheim, Germany

Jiménez-González C, Curzons AD, Constable DJC et al (2004) Cradle-to-gate life cycle inventory and assessment of pharmaceutical compounds. Int J Life Cycle Assess 9:114–121

Jiménez-González C, Poechlauer P, Broxterman QB et al (2011) Key green engineering research areas for sustainable manufacturing: a perspective from pharmaceutical and fine chemicals manufacturers. Org Process Res Dev 15:900–911

Jiménez-González C, Ollech C, Pyrz W et al (2013) Expanding the boundaries: developing a streamlined tool for eco-footprinting of pharmaceuticals. Org Process Res Dev 17:239–246

Jolliet O, Margni M, Charles R et al (2003) Impact 2002+: a new life cycle impact assessment methodology. Int J Life Cycle Assess 6:324–330

Kralisch D, Ott D, Gericke D (2015) Rules and benefits of life cycle assessment in green chemical process and synthesis design: a tutorial review. Green Chem 17:123–145

Lanzafame P, Centi G, Perathoner S (2014) Evolving scenarios for biorefineries and the impact on catalysis. Catal Today 234:2–12

Lapkin A, Constable DJC (2009) Green Chemistry metrics: measuring and monitoring sustainable processes. © 2009 Blackwell Publishing Ltd. ISBN 978-1-405-15968-5

Lee E, Andrews CJ, Anctil A (2018) An iterative approach to evaluate and guide fine chemical processes: an example from chloroaluminum phthalocyanine for photovoltaic applications. ACS Sustain Chem Eng 6:8230–8237

Leseurre L, Merea C, Duprat de Paule S et al (2014) Eco-footprint: a new tool for the "Made in Chimex" considered approach. Green Chem 16:1139–1148

Lichtlen L, Pochard C (2018) Danone, Nestlé waters and origin materials welcome PepsiCo to the NaturALL Bottle Alliance. https://www.nestle-waters.com/asset-library/documents/press%20releases/2018/pepsico_joins_naturall_alliance.pdf. Accessed 28 Feb 2019

Marion P, Bernela B, Piccirilli A et al (2017) Sustainable chemistry: how to produce better from less? Green Chem 19:4973-4989

Merck KGaA (2019). http://www.merckmillipore.com/IT/it/responsibility/products/design-innovation/Life-Cycle-Assessment/pBeb.qB.U58AAAFD8ldZXujz,nav. Accessed 28 Feb 2019

Minoux D, Nesterenko N, Vermeiren W et al (2009) WO Patent 2009098268A1, Dehydration of alcohols in the presence of an inert component. Total Petrochemicals Research Feluy

Muñoz I (2012) LCA in green chemistry: a new subject area and call for papers. Int J Life Cycle Ass 17:517–519

Novamont (2019). https://www.novamont.com/eng/. Accessed 28 Feb 2019

P&G—The Procter & Gamble Company (2019). https://us.pg.com/environmental-sustainability/. Accessed 28 Feb 2019

Patel AD, Meesters K, Den Uil H et al (2012) Sustainability assessment of novel chemical processes at early stage: application to biobased processes. Energ Environ Sci 5:8430–8444

Patrick Gracey B, Partington SR (2009) Process for preparing ethene. WO Patent 2009050433A1, BP plc

Phan TVT, Gallardo C, Mane J (2015) GREEN MOTION: a new and easy to use green chemistry metric from laboratories to industry. Green Chem 17:2846–2852

Philippe M, Didillon B, Gilbert L (2012) Industrial commitment to green and sustainable chemistry: using renewable materials & developing eco-friendly processes and ingredients in cosmetics. Green Chem 14:952–956

PRé Consultants (2019) Fact-based sustainable development. https://www.pre-sustainability.com/. Accessed 28 Feb 2019
Rennovia (2019). http://www.rennovia.com/. Accessed 28 Feb 2019
Ricciardi M, Cespi D, Celentano M et al (2017a) Bio-propylene glycol as value added product from Epicerol® process. Sustain Chem Pharm 6:10–13
Ricciardi M, Passarini F, Vassura I et al (2017b) Glycidol, a valuable substrate for the synthesis of Monoalkyl Glyceryl Ethers: a simplified life cycle approach. Chemsuschem 10:2291–2300
Ricciardi M, Passarini F, Capacchione C et al (2018) First attempt of Glycidol-to-Monoalkyl Glyceryl Ethers conversion by acid heterogeneous catalysis: synthesis and simplified sustainability assessment. Chemsuschem 11:1829–1837
Roschangar F, Zhou Y, Constable DJC et al (2018) Inspiring process innovation via an improved green manufacturing metric. iGALGreen Chem 20:2206–2211
Ruiz-Mercado GJ, Smith RL, Gonzalez MA (2012a) Sustainability indicators for chemical processes: I. Taxonomy. Ind Eng Chem Res 51:2309–2328
Ruiz-Mercado GJ, Smith RL, Gonzalez MA (2012b) Sustainability indicators for chemical processes: II. Data Needs. Ind Eng Chem Res 51:2329–2353
Ruiz-Mercado GJ, Carvalho A, Cabezas H (2016) Using Green Chemistry and engineering principles to design, assess, and retrofit chemical processes for sustainability. ACS Sustain Chem Eng 4:6208–6221
Saling P, Kicherer A, Dittrich-Krãmer B et al (2002) Eco-efficiency analysis by BASF: the method. Int J Life Cycle Assess 4:203–218
Saling P, Maisch R, Silvani M et al (2005) Assessing the environmental-hazard potential for life cycle assessment, eco-efficiency and SEEbalance. Int J Life Cycle Assess 10:364–371
Salvadori R (1930) Merceologia generale. Poligrafica universitaria, Firenze, Italy
Schmidt I, Meurer M, Saling P et al (2004) Managing sustainability of products and processes with the socio-eco-efficiency analysis by BASF. GMI 45 Spring. http://www.academia.edu/25480624/Managing_sustainability_of_products_and_processes_with_the_socio-eco-efficiency_analysis_by_BASF. Accessed 28 Feb 2019
SDGs. https://sustainabledevelopment.un.org/?menu=1300. Accessed 28 Feb 2019
Secchi M, Castellani V, Collina E et al (2016) Assessing eco-innovations in green chemistry: Life Cycle Assessment (LCA) of a cosmetic product with a bio-based ingredient. J Clean Prod 129:269–281
SGI. Sustainable guar iniziative. https://www.solvay.com/en/sustainability/acting-society/sustainable-guar-initiative. Accessed on 28 Feb 2019
Sheldon RA (1992) Organic synthesis-past, present and future. Chem Ind 23:903–906
Sheldon RA (1997) Catalysis and pollution prevention. Chem Ind 1:12–15
Sheldon RA (2018) Metrics of Green Chemistry and sustainability: past, present, and future. ACS Sustain Chem Eng 6:32–48
Shonnard A, Kicherer A, Saling P (2003) Saling, industrial application using BASF eco-efficiency analysis: perspectives on green engineering principles. Environ Sci Technol 37:5340–5348
Smith H (1968) The cumulative energy requirements of some final products of the chemical industry. In: World power conference, Moscow, Russia
Solvay S.A. (2019). https://www.solvay.com/en/sustainability/acting-sustainable-business/sustainable-portfolio-management-spm-tool. Accessed 28 Feb 2019
Subramaniam E, Semenzin A, Zabeo P et al (2018) Assessing the social impacts of anno-enabled products through the life cycle: the case of nano-enabled biocidal paint. Int J Life Cycle Assess 2:348–356
Swarr TE, Cucciniello R, Cespi D (2019) Environmental certifications and programs roadmap for a sustainable chemical industry. Green Chem 21:375–380
Traverso M, Bell L, Saling P, Fontes J (2018) Towards social life cycle assessment: a quantitative product social impact assessment. Int J Life Cycle Assess 3:597–606
Tripodi A, Bahadori E, Cespi D et al (2017) Acetonitrile from bioethanol ammoxidation: process design from the grass-roots and life cycle analysis. ACS Sustain Chem Eng 6:5441–5451

Trost BM (1991) The atom economy—a search for synthetic efficiency. Science 254:1471–1477
UNEP, United Nation Environmental Programme (2013) The Methodological Sheets for Sub-categories in Social Life Cycle Assessment (S-LCA)
UNEP/SETAC (2009) Guidelines for social life cycle assessment of products. United Nations Environmental Porgram, Paris SETAC Life Cycle Initiative United Nation Environmental Program
Van Schoubroeck S, Van Dael M, Van Passel S, Int J of Life Cycle Assess (2018) A review of sustainability indicators for biobased chemicals. Renew Sust Energ Rev 94:115–126
Vermeiren W, Van Gyseghem N (2011) A process for the production of bio-naphtha from complex mixtures of natural occurring fats & oils. WO Patent 2011012439A1, Total Petrochemicals Research Feluy
Weiss M, Haufe J, Carus M et al (2012) A review of the environmental impacts of biobased materials. J Ind Ecol 16:S169–S181
World Bank Group (2016) ECOFYS. Carbon Pricing Watch 2016. Washington, DC, USA. https://openknowledge.worldbank.org/handle/10986/24288. Accessed 28 Feb 2019

4 The LCA Modelling of Chemical Companies in the Industrial Symbiosis Perspective: Allocation Approaches and Regulatory Framework

Emanuela Viganò, Carlo Brondi, Simone Cornago, Antonio Caretta, Letizia Bua, Lino Carnelli, Giovanni Dotelli, Michael Martin and Andrea Ballarino

Abstract

The application of industrial symbiosis to chemical processes is made possible by considering physical by-products as new sources in spite of potential waste. The by-product acts as a substitution of raw materials in symbiotic chemical companies, possibly taking into account geographical proximity. While advantages appear evident in terms of reduced cost savings and resource consumption, benefits related to other environmental impacts seem to be less clear. This chapter deals with different case studies of industrial symbiosis and how these are treated according to literature, regulatory approaches and main guidelines. Once the different approaches and their range of applications are presented, the chapter describes the main barriers and strengths in the application of these rules to the chemical sector. Finally, the identified approaches are applied to a specific case of industrial symbiosis in the chemical sector by illustrating how the single data can be calculated and can quantitatively change.

E. Viganò (✉)
CESI S.p.A., Milan, Italy
e-mail: emanuela.vigano@cesi.it

C. Brondi · S. Cornago · A. Ballarino
CNR-STIIMA National Research Council, Milan, Italy

A. Caretta · L. Bua · L. Carnelli
ENI S.p.A. Renewable Energy and Environmental R&D, Novara, Italy

G. Dotelli
Department of Chemistry, Materials and Chemical Engineering "G. Natta", Politecnico di Milano, Milan, Italy

M. Martin
IVL Swedish Environmental Research Institute, Stockholm, Sweden

S. Maranghi and C. Brondi (eds.), *Life Cycle Assessment in the Chemical Product Chain*, https://doi.org/10.1007/978-3-030-34424-5_4

This chapter aims to guide the LCA practitioner through an agile set of rules in modelling industrial symbiosis in an LCA perspective and to provide a quantitative evaluation of the effects that a modelling choice may produce with respect to another.

4.1 Introduction

Industrial Symbiosis (IS) is a key sub-concept of industrial ecology since its initial formulation (Baldassarre et al. 2019). IS represents a collaborative approach in which different industries create a cooperative network to exchange materials, energy, water and/or by-products. IS can play a prominent role in the change towards sustainable development (Baldassarre et al. 2019). In fact, local cooperation in industrial symbiosis can reduce raw materials use and waste disposal, but material and energy flows extending outside symbiosis boundaries involve environmental effects by addressing issues related to resource depletion, waste management and pollution Blengini et al. (2012). These external impacts risk being ignored in industrial symbiosis studies that focus just on local environmental aspects, on the contrary, additional effects due to substitution can introduce relevant effects outside the boundaries of the symbiotic network. In general, LCA approaches allow identifying areas in which industrial symbiosis and sustainability can proficiently work. Beyond the assessment of the environmental burden, a rising area of interest is represented by rules revision in order to properly attribute environmental effects (Baldassarre et al. 2019). Such an approach can support the shift from assessing industrial ecology aspects to proper design of industrial symbiosis within a multiple product chain (Curran and Williams 2012; Daddi et al. 2017; Dong et al. 2014; Huang et al. 2019; Kerdlap et al. 2019; Liu et al. 2011; Pushkar 2019).

In terms of LCA debate, the case of industrial symbiosis can represent a relevant issue in the calculation methods from different points of view. In fact, the flow accounting implies that contextual matter and energy flows are by-products affecting different product life cycles. From the perspective of an isolated life cycle, an additional shift of waste into by-product is avoided waste Dong et al. (2017). From the perspective of a complementary life cycle the additional by-product represents a certain substitution of raw resource. At the same time, by-product involves additional treatments and transportation for a proper substitution with their related additional environmental impacts, these further impacts still requiring to be subdivided among the entering and leaving life cycle. The emerging LCA issue is then how to distribute additional impact due to this type of flow within the LCA. The concurrence of different rules involves difficulties in flow attribution according to the LCA aims and the general framework in which a rule is applied (e.g. PEF or EPD labelling). This chapter aims to examine the existing rules on allocation in the case of industrial symbiosis. In particular, the different rules of allocation and displacement of the system boundaries are listed and examined in order to understand how the calculation changes. Precisely, a specific case of symbiosis is examined: a by-product is used to produce energy; the effects of the chosen criteria according to the variation of the CO_2 calculation are observed Ammenberg et al. (2015). Finally, in the part of the discussion a series of emerging suggestions are elaborated.

4.2 Overview on Literature for Allocation in the Chemical Sector

Life Cycle Assessment (LCA) has been applied widely in the academic literature for the quantification of the environmental performance of industrial symbiosis (IS) networks (Martin 2015; Sokka et al. 2011; Røyne et al. 2018; Martin and Harris 2018; Chertow and Lombardi; Brondi et al. 2018). Many of these studies have employed LCA using different life cycle based techniques and varying scopes.

Currently, different approaches can be concurrently adopted in assessing environmental effects of substitution within circular networks. On the other hand, in the absence of agreed rules, methods and indicators, further steps of optimization and efficiency of industrial symbiosis can have limited effects and comparisons (Brondi et al. 2018)

Nonetheless, there is a divergent set of recommendations outlined in the approaches and in the methodological guidelines among scholars in the field. Methodological guidance for IS environmental performance quantification is provided in several studies, such as Martin et al. (2015), Mattila et al. (2011) and Kim et al. (2017). These studies recognize many similarities, suggesting that the quantifications are sensitive to a number of methodological choices. Of utmost importance is the choice of the reference system, the system boundaries, the allocation method (or lack thereof in the case of system expansion techniques) and the functional unit employed for comparisons; see e.g. approaches outlined in (Martin et al. 2015; Kim et al. 2017; Mattila et al. 2012; Sokka et al. 2011; Seto et al. 2017). As addressed by Kim et al. (2017) and Røyne et al. (2018) the results can also be sensitive to the modelling, and ultimately the system boundaries, where failing to review upstream and downstream impacts, may have significant implications for the sustainability of the system.

While many industrial sectors taking part in IS networks are not influenced by LCA guidelines, the biofuel and bioenergy industry is a relevant industrial sector to have LCA as mandatory from policy (RED II 2018). Martin (2015) and Lazarevic and Martin (2018) and Martin and Harris (2018), review the implications of policy influence on the use and the quantitative assessments of LCAs when biofuels are a product of an IS network. As such, the guidelines provided in the academic literature may vary significantly according to the allocation rules and methods provided by the Renewable Energy Directive (RED), which applies a combination of attributional and consequential approaches for reviewing the biofuels impact and the integration with other systems (e.g. district heating and carbon dioxide storage systems, Ekvall and Finnveden 2001). A second set of policy rules is provided by the PEF (product environmental footprint). Also in this case, the application of a set of mediated rules reporting substitution effects on a larger scale by merging both consequential and attributional approaches can fail in assessing the effective environmental impact of substitution in the chemical sector. As first conclusion, the main drivers in the determination of LCA results in the case of industrial symbiosis strictly depend on allocation choices (Mohammed et al. 2016 and 2018). Different drivers can be further identified and more deeply presented.

- *Allocation type*: procedures for allocation depend on the application sectors as well as the type of available data. Traditional mass and cost allocation can be supported by other chemical-specific allocations like volumetric allocation, energy or molar content.
- *System boundaries and point of substitution for allocation*: definition of system boundaries and product life cycle include the understanding of effective reuse of by-products. Such a challenge implies to identify complementary life cycles interacting with the expected by-product. The case of multiple reuse of the same by-product can create difficulties in the proper interpretation of results.
- *Assumption on by-products*: identification of by-products reuse includes the proper definition of possible reuse and chemical properties. Effective reuse firstly implies identifying a transition from waste to a specific substitution process, which means that transport and treatment processes should be identified. Secondly, under the same hypothesis a by-product should proficiently substitute a specific raw material for the same functional profile.
- *Technical and economic scenarios*: The allocation of the impacts strictly depends on how effectively a raw material can be substituted by secondary flow. In such a term, both economic and technical figures compete in determining such efficiency. Technical figures mirror the transformation process efficiency due both to the pre-treatment process and to direct application within the final product. Economic figures represent market substitution in terms of value for secondary streams.

4.2.1 Allocation Procedures for Industrial Symbiosis

Industrial symbiosis in the chemical sector requires a particular configuration of product lifecycles requiring that a by-product is used as processing material. Main allocation procedures follow general literature and legal prescriptions. Hereafter main outcomes from literature, policy and legal guidelines will be reported. In terms of literature review, an extensive overview is presented with reference to Martin et al. (2015) work for LCA to IS networks for comparisons. As far as the general policy guidelines—the RED guideline (RED II 2018)—the case of by-products oriented to energy sector is presented. The RED method is used also in cases—as the one presented hereafter—which involve biofuel industry, with biodiesel as a main product. A brief reference to allocation in PEF guideline is also provided.

4.2.2 Renewable Energy Directive

The European Commission within the 'Clean Energy for all Europeans' initiative has adopted the Renewable Energy Directive (RED). The Renewable Energy Directive (2009/28/EC) established a European legal framework for the promotion of renewable energy as a relevant part of the European internal consumption. In particular, RED set mandatory national renewable energy targets in order to achieve a target of an overall 20% share of renewable energy in the EU energy consumption by 2020. After the first release in November 2016, the commission released a new version (RED II) in late 2018 with higher targets (Renewable Energy Sources consumption by 2030 has been raised to 32%). The RED II sets up also a number of sub-target in different areas as public transport, bio-sources features, use of natural stocks, etc. In terms of energy for circular economy, the RED II fixes the main limits for energy from bio-sources in order both to support the raise of quantity of bioenergy from the current 8.8% and to fulfil limits for carbon emissions. The RED II further extends sustainability criteria to biomass fuels used for power, heating and cooling production, particularly for larger installations.

For forest biomass, a risk-based approach is applied in order to minimize the possibility of pattern forest biomass derived from unsustainable production and alignment with land use and land use modification. RED II establishes also a proper carbon accounting. Suppliers of bioenergy have then to accommodate the threshold value for biofuels and bioenergy in order to be compliant with the reference level.

RED II states rules for calculating the greenhouse gas impact of biofuels, bio-liquids and their fossil fuel comparators as well as greenhouse gas emissions saving default value as g CO_2eq/MJ.

In the specific study case greenhouse gas emissions from the production and use of biomass fuels before conversion into electricity, heating and cooling shall be calculated as in (4.1) according to the RED parameters listed below:

$$E = e_{ec} + e_{l} + e_{p} + e_{td} + e_{u} - e_{sca} - e_{ccs} - e_{ccr} \quad (4.1)$$

Emissions from the manufacture of machinery and equipment shall not be taken into account. RED II similarly provides allocation rules for the calculation of transportation fuels from renewable sources. However, this kind of biosource is not included in the case of the industrial symbiosis but as a single product in the transport sector (Table 4.1).

4.2.3 Circular Footprint Formula (CFF)

The PEF initiative is a European initiative set up in the year 2012 to establish a common methodological approach to enable Member States and the private sector to assess, display and benchmark the environmental performance of products, services and companies based on an LCA-based assessment over the product

Table 4.1 Parameters in the greenhouse gas emission calculation

Terms	Explanation
E	Total emissions from the production of the fuel before energy conversion
e_{ec}	Emissions from the extraction or cultivation of raw materials
e_l	Annualized emissions from carbon stock changes caused by land use change
e_p	Emissions from processing
e_{td}	Emissions from transport and distribution
e_u	Emissions from the fuel in use
e_{sca}	Emission savings from soil carbon accumulation via agricultural management
e_{ccs}	Emission savings from CO_2 capture and geological storage
e_{ccr}	Emission savings from CO_2 capture and replacement

lifecycle (Zampori et al. 2019). Contribution in terms of Industrial Symbiosis (IS) is mirrored in the environmental profile calculation in two phases.

- In terms of raw material acquisition, the pre-processing of all material inputs to the studied product are included and the assessment is limited to an attributional perspective. Then the pre-processing and transportation of recycled materials from other industries is entirely attributed to the new product life cycle.
- In terms of the end-of-life stage, the impact shall be modelled using the Circular Footprint Formula (CFF). Such a formula partially includes an attributional and a consequential perspective by representing the multiple impacts and substitution for material energy and disposal.

More in particular the Circular Footprint Formula (CFF) is a combination of material (4.2) + energy (4.3) + disposal (4.2), (4.3), (4.4).

CFF for material

$$(1-R_1)E_V + R_1 \times (\mathrm{AE}_{\mathrm{recycled}} + (1-A)E_V \times \mathrm{Qs}_{\mathrm{in}}/Qp) + (1-A)\,R_2 \times (E_{\mathrm{recyclingEoL}} - E_V^* \times \mathrm{Qs}_{\mathrm{out}}/\mathrm{QP}) \tag{4.2}$$

CFF for energy

$$(1-B)R_3 \times (E_{ER} - \mathrm{LHV} \times X_{\mathrm{ER,heat}} \times E_{\mathrm{SE,heat}} - \mathrm{LHV} \times X_{\mathrm{ER,elec}} \times E_{\mathrm{SE,elec}}) \tag{4.3}$$

CFF for disposal

$$(1-R_2-R_3) \times E_D \tag{4.4}$$

In a cradle-to-gate PEF study the parameters (Table 4.2) related to the end of life of the product shall not be calculated.

Table 4.2 Parameters related to Industrial Symbiosis in CFF

Terms	Explanation
A	Allocation factor of burdens and credits between players of recycled materials
B	Allocation factor of processes involving energy recovery. It is applied both to burdens and credits
Qs_{in}	Quality of the entering secondary material, i.e. the quality of the recycled material at the point of substitution
Qs_{out}	Quality of the produced secondary material, i.e. the quality of the recyclable material at the point of substitution
Q_p	Quality of the primary material, i.e. quality of the virgin material
R_1	Proportion of material in the input to the production that has been recycled from a previous system
R_2	Proportion of the material in the product that will be recycled (or reused) in a subsequent system. R_2 shall therefore take into account the inefficiencies in the collection and recycling (or reuse) processes. R_2 shall be measured at the output of the recycling plant
R_3	Proportion of the material in the product that is used for energy recovery at EoL
$E_{recycled}$ (E_{rec})	Specific emissions and resources consumed (per functional unit) arising from the recycling process of the recycled (reused) material, including collection, sorting and transportation process
$E_{recyclingEoL}$ (E_{recEoL})	Specific emissions and resources consumed (per functional unit) arising from the recycling process at EoL, including collection, sorting and transportation process
E_v	Specific emissions and resources consumed (per functional unit) arising from the acquisition and pre-processing of virgin material
E_v^*	Specific emissions and resources consumed (per functional unit) arising from the acquisition and pre-processing of virgin material assumed to be substituted by recyclable materials
E_{ER}	Specific emissions and resources consumed (per functional unit) arising from the energy recovery process (e.g. incineration with energy recovery, landfill with energy recovery, etc.)
$E_{SE,eat}$ $E_{SE,lec}$	Specific emissions and resources consumed (per functional unit) that would have arisen from the specific substituted energy source, heat and electricity, respectively
E_D	Specific emissions and resources consumed (per functional unit) arising from disposal of waste material at the EoL of the analysed product, without energy recovery
$X_{ER,heat}$ $X_{ER,elec}$	The efficiency of the energy recovery process for both heat and electricity
LHV	Lower heating value of the material in the product that is used for energy recovery

The CFF including different criteria to properly report substitution within different life cycles, users of the PEF method shall report all the parameters used. Some of them come from specific assumption on industrial symbiosis even if it is not explicitly cited within the guideline. On the other hand, other allocation aspects depend not simply on factors but on methodological choices (i.e. point of substitution):

- *The A factor*: such a factor is related to industrial symbiosis in case the by-product is directly sold on the market. The A factor allocates burdens and credits from recycling and virgin material production between two life cycles (i.e. the one supplying and the one using recycled material) (Chen et al. 2010) and it aims to reflect market realities. An A factor equal to 1 would reflect a 100:0 approach (i.e. credits are given to the recycled content), an A factor equal to 0 would reflect a 0:100 approach (i.e. credits are given to the recyclable materials at the end of life). In PEF studies the A factor values shall be in the range $0.2 \leq A \leq 0.8$, to always capture both aspects of recycling (recycled content and recyclability at end of life). The driver to determine the values of the A factor is the analysis of the market situation from a low offer situation ($A = 0.2$) to a market equilibrium between offer and demand ($A = 0.5$). In the case a material-specific A value is not available, the A value shall be set equal to 0.5.
- *The B factor*: the B factor represents the allocation factor for the energy recovery processes in case material. It applies both to burdens and credits. Credits refer to the amount of heat and electricity sold, not to the total produced, taking into account relevant variations over a 12-month period, e.g. for heat. In PEF studies the B value shall be equal to 0 as default. To avoid double-counting between the current and the subsequent system in case of energy recovery, the subsequent system shall model its own energy use as primary energy.
- *The point of substitution*: allocation is addressed by the determination of the point of substitution in order to apply the material part of the CFF. The point of substitution corresponds to the point in the value chain where secondary materials substitute primary materials. The point of substitution shall be identified in correspondence to the process where input flows are coming from 100% primary sources and 100% secondary sources. In some cases, the point of substitution may be identified after some mixing of primary and secondary material flows has occurred.
- *The quality ratios* Qs_{in}/Q_p *and* Qs_{out}/Q_p: in order to address the proper substitution of raw material with a by-product two ratios can be used in the CFF, to assess the quality of both the ingoing and the outgoing recycled materials. The assessment of the quality ratios has to be based both on economic aspects (the price of secondary materials in comparison to primary resources at the point of substitution) and on physical aspects (in case economic aspects are less relevant than physical aspect). It is important to remark that in case $E_v = E_v^*$, the two quality ratios are needed: Qs_{in}/Q_p that are linked to the recycled content and Qs_{out}/Q_p that are associated to recyclability at EoL. These factors are added to capture the downcycling of material with reference to the original primary material and can also capture the effect of multiple recycling processes.
- *The recycled content (R_1)*: such a factor is referred to the real quantity that can be obtained due to industrial symbiotic processes. The company has to provide evidence about the percentage of recycled input material into the resulting end

products through its management system. In fact, material-specific values based on supply market statistics cannot be accepted as a proxy. The R_1 values applied shall result from supply-chain- or application-specific assessments, depending on the information accessible by the company performing the PEF study.

- *The factors* $E_{recycled}$ *(*E_{rec}*),* $E_{recyclingEoL}$ *(*E_{recEoL}*)*: the system boundary of *E*rec and ErecEoL has to consider all the produced emissions and the consumption of resources from the collection up to the defined point of substitution. In case the point of substitution is identified directly for the material to be included in the product (level 2) *E*rec and *E*recEoL have to be modelled on the basis of the actual entering flows. Therefore, in case a portion of the input flows comes from raw materials, they have to be included within the datasets that are used to model *E*rec and E_{recEoL}. In specific cases, E_{rec} may correspond to E_{recEoL}, as in the cases of the closed loops.
- *The* E_v^*: when default E_v^* equals E_v, the user shall assume that a recyclable material at end-of-life substitutes the same virgin material which was used as input to produce that material. In some cases, E_v^* will be different from E_v. In this case, the user shall provide evidence that a recyclable material is substituting a different virgin material than the one producing the recyclable material. In case $E_v^* \neq E_v$, E_v^* corresponds to the amount of virgin material that can be replaced by the recyclable material. In such cases, E_v^* is not multiplied by Qs_{out}/Q_p, in fact such parameter is already taken into account when calculating the 'actual amount' of virgin material substituted. Such quantity shall be calculated taking into account that the virgin material substituted and the recyclable material fulfil the same function in terms of 'how long' and 'how well'. E_v^* shall be determined by considering evidence of real substitution of the specified virgin materials.

In case a waste material is reused as a fuel, it shall be considered as an energy recovery process within the 'energy' part of CFF equations.

4.3 Life Cycle Based IS Method

As outlined in Martin et al. (2015) the selection, and potential impacts, of methodological considerations used for the quantifications including, e.g. the choice of reference systems, allocation methods, system boundaries and functional unit(s) are imperative for the system. Figure 4.1 outlines the system boundaries of the industrial symbiosis network, the functional units, the selection of by-products and avoided products included in the system.

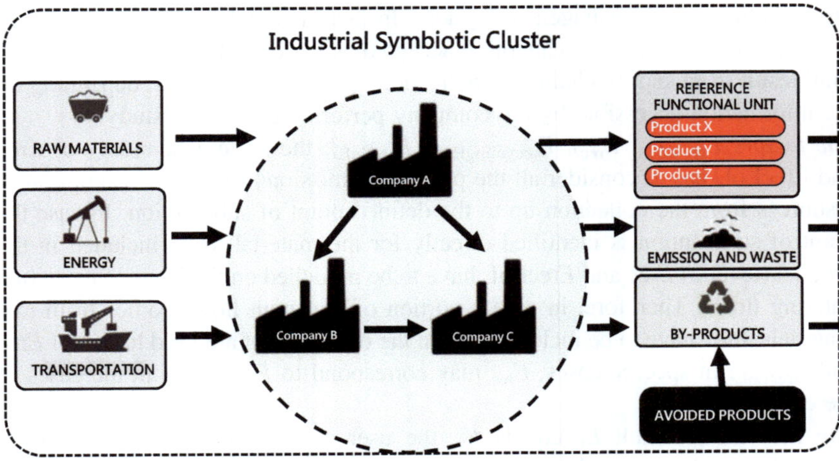

Fig. 4.1 Illustration of methodological choices and data for reviewing an IS network (*Source* Martin et al. 2015)

4.3.1 50/50 Method

Figure 4.2 illustrates the approach for a synergy between Firms A and B. Using the 50/50 method outlined in Martin et al. (2015), the distribution of credits from the avoidance of Raw B is accounted for by providing Firm A and B a share (50% each) of the credit for the equivalent amount of Raw B avoided. In order to not double-count the potential benefits by removal of Raw B (and the change in its use), Firm B is provided a burden/impact for the production of Raw B; thus Firm B would only receive 50% of the impact of Raw B in total. Furthermore, by-products leaving the system are still avoided, according to the use of system expansion methodology. Intermediate processes may also be required for the use of By-product A in the processing for Firm B, and the impacts of such a step are to be distributed between the firms involved in the exchange following the same 50/50 logic; this can include distributing impacts from upgrading and transport between the different companies involved in an exchange. Extending this method, according to the specific cases, processes, by-products and allocation methods considered, this percentage distribution could be varied from the 50–50 method outlined in Martin et al. (2015)—sharing equally the credits of the avoided impacts to the two subsystems involved—to other percentages distribution, assuming a 100% total sum and arriving to the extreme 100–0 and 0–100 cases considered in the case study modelling as limit cases. The 50–50 method enables equal and impartial benefits (e.g. credits) distribution without needing sensible considerations on different worth and responsibilities in the process. In practical applications, the limit percentages should be adopted only under justified conditions. Other intermediate conditions (75–25, 25–75) could be adopted in case of different involvements of the involved subjects/subsystems, but always providing motivations.

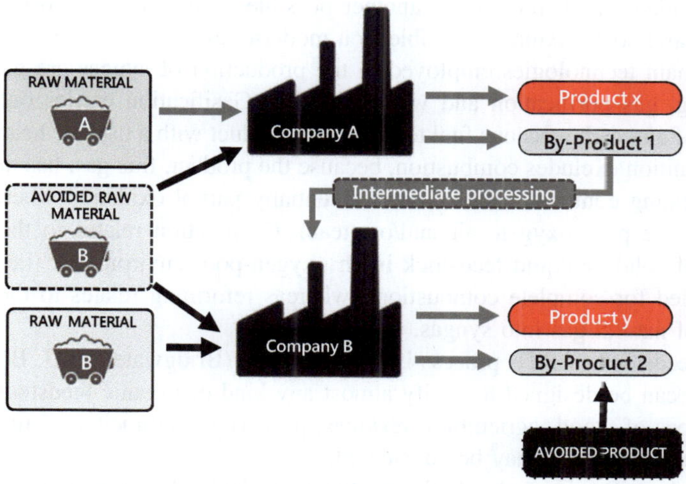

Fig. 4.2 Illustration of the methodology using 50/50 method (*Source* Martin et al. 2015)

4.4 Case Study

The case study aims to provide general suggestions to practitioners in order to model cases of industrial symbiosis with different allocation perspectives. The case study is based on a real industrial case; the model introduces different degrees in symbiosis and in allocation between different players.

4.4.1 Bio-Refinery Plant Symbiotic Flows

The case study concerns a bio-refinery plant, whose symbiotic flow descriptions are hereunder explained.

The Biomass-to-Liquids process (BTL) is a thermo-catalytic pathway to transform biomass to liquid fuels, in particular to middle distillates including diesel cuts.

The main steps of this technology include thermal gasification of biomass, followed by syngas clean up, Fischer—Tropsch synthesis and hydrocarbon isomerization to give suitable synthetic fuels (Perego et al. 2010; Zennaro et al. 2013).

Similar processes are well known and currently used to produce synthetic fuels from coal or gas (i.e. CTL and GTL). In fact, when syngas is used as feedstock in the Fischer—Tropsch (FT) synthesis, the overall process is generally named XTL, X, depending on the carbon source, for example, CTL coal-to-liquids, GTL natural gas-to-liquids, BTL biomass-to-liquids, or WTL waste-to-liquids (Zennaro et al. 2013). What changes is the feedstock to gasification. Combining coal and biomass, in a so-called CBTL process, is another possible route since the co-feeding of a biomass and coal mixture is feasible in a modern gasifier.

The main technologies employed in the production of syngas are gasification, reforming, gas purification and water-gas shift. Gasification carries out the conversion of any carbonaceous fuel to a gaseous product with a useable heating value. This definition excludes combustion, because the product, flue gas, has no longer a useful heating content. Gasification is essentially partial oxidation where the oxidant may be pure oxygen, air and/or steam. Gasification relates to the transformation of solid or liquid feedstock in an oxygen-poor environment (i.e. less than that needed for complete combustion), whereas reforming relates to the transformation of natural gas into syngas.

The feedstock of BTL process is solid biomass (Bridgwater 2003; Basu 2006). Gasifiers can be designed to gasify almost any kind of organic feedstock, such as many types of wood, agricultural residues, peat, coal, anthracite, oil residues and municipal solid waste may be considered.

Several gasification technologies are today available, differing on the feedstock's specifications (size, chemical composition), nevertheless the reactor type (fixed or moving) and the heating method are the main elements characterizing the families of gasifiers. These are grouped into four main technologies: fixed bed, fluidized bed (bubbling or circulating), entrained flow and indirect.

In the present chapter a direct fluidized bed gasifier (Chiodini et al. 2017) using oxygen and steam as gasifying agents has been considered. Since gasification needs an inlet flow as dry as possible, a biomass pretreatment unit is needed to reduce the water content from about 50–30% to about 15% wt.

After the gasification unit, the produced syngas is purified in a cleaning gas unit to eliminate dust, tars and water-soluble gas like hydrogen chloride and ammonia. Reforming and water-gas shift, WGS, sections adjust the H_2 to CO ratio in order to obtain the correct stoichiometric one for Fischer—Tropsch synthesis, approximately 2 (de Souza-Santos 1989). Reforming takes place in a catalytic steam methane reformer (SMR), the kinetic of which is controlled by a nickel catalyst (Ji et al. 2009; Jess 1996), also off-gas C1–C4 coming from the FT reactor can be fed to this. The water-gas shift reaction, which transforms carbon monoxide and water into carbonic anhydride and hydrogen again increasing the H_2/CO ratio, is a moderately exothermic reversible reaction and preferably takes place at 200–250 °C on a Cu–Zn–Al catalyst.

Afterwards, the syngas is sent to the Rectisol® island for the acid gases (CO_2 and H_2S) removal (Rank 1972). The syngas treatments, after the gasification step, are necessary for different reasons: the correct H_2 to CO ratio is required for FT catalysis, the sulphur compounds are poison for the catalyst of the FT and the CO_2 contributes to lower the reagents partial pressure as an inert gas in the FT reaction (Tijimensen et al. 2002). Fischer—Tropsch reactions produce hydrocarbons starting from H_2 and CO (syngas) and can take place in a Slurry Bubble Column Reactor (SBCR) or in a fixed bed reactor on cobalt or iron catalysts.

The Fischer—Tropsch reaction produces long linear waxes not directly usable as fuels so a final process step of upgrading (UPG) of the products to produce diesel

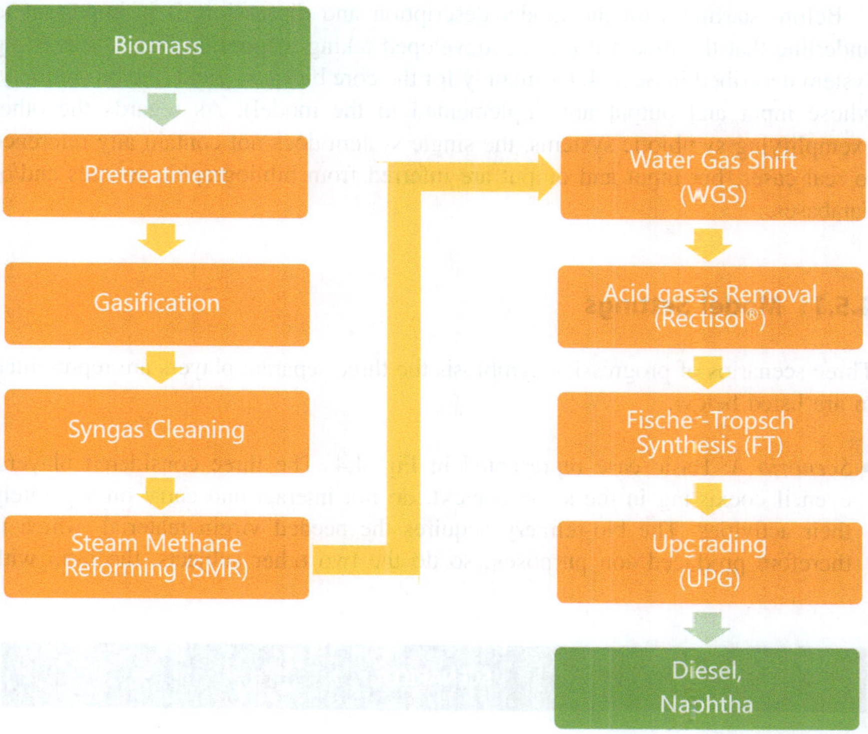

Fig. 4.3 BTL process flow diagram

and naphtha is necessary. It is made by hydrocracking and isomerization reactions. A final distillation section separates the products in the commercial fractions, naphtha, jet fuel, diesel, and recycle light and heavy tails. Figure 4.3 shows the overall BTL multistep path.

4.5 Modelling

The modelling of the case study is carried out by adopting in this case a single allocation model between complementary systems. In particular, three players are identified in a reduced network: a bio-refinery (System 2), a farm (System 1) and an energy provider (naphtha user, System 3). The bio-refinery exploits some virgin biomass to produce biodiesel and bio-naphtha; the considered farm starting from seeds produces the related crop and a certain amount of waste biomass; a chemicals producer uses some petroleum-based naphtha to produce ethylene.

Before starting with the model description and discussion, it is important to underline that the model itself was developed taking inspiration from the existing system described in Sect. 4.4.1, mainly for the core business yield (the bio-refinery, whose input and output are implemented in the model). As regards the other exemplifying symbiotic systems, the single system does not contain any reference to real cases, but input and output are inferred from bibliographic sources and/or databases.

4.5.1 Model Settings

Three scenarios of progressive symbiosis the three separate players are represented in are listed below.

- *Scenario A*: Basic case represented in Fig. 4.4. The three considered players, even if coexisting in the same context, do not interact and carry on separately their activities. The bio-refinery acquires the needed virgin material, which is therefore produced «on purpose», so do the two other subjects, the farm with

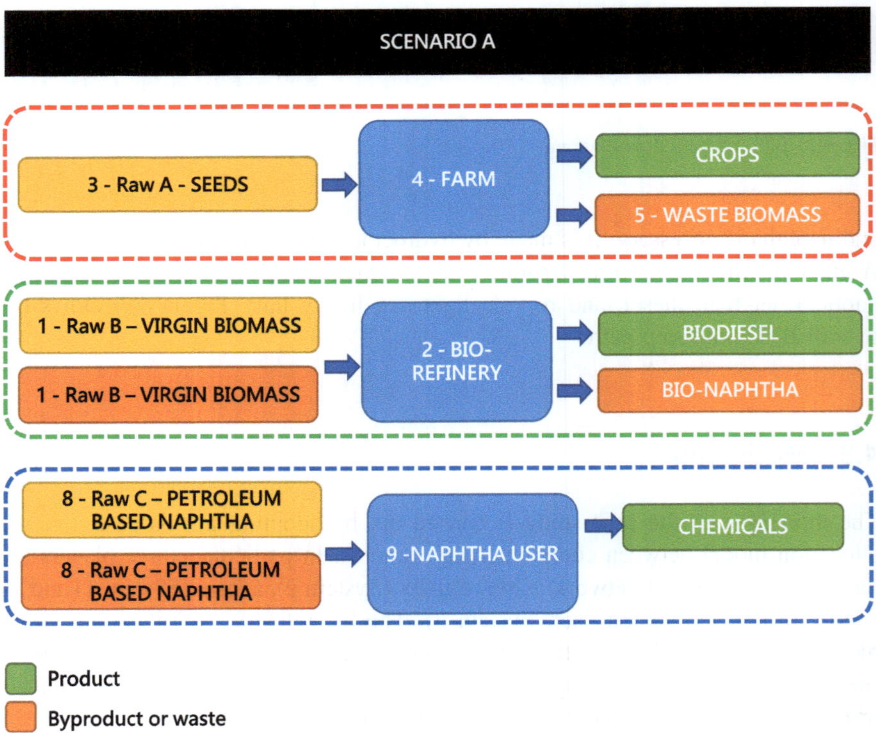

Fig. 4.4 Scenario 1, Scenario A three separated systems

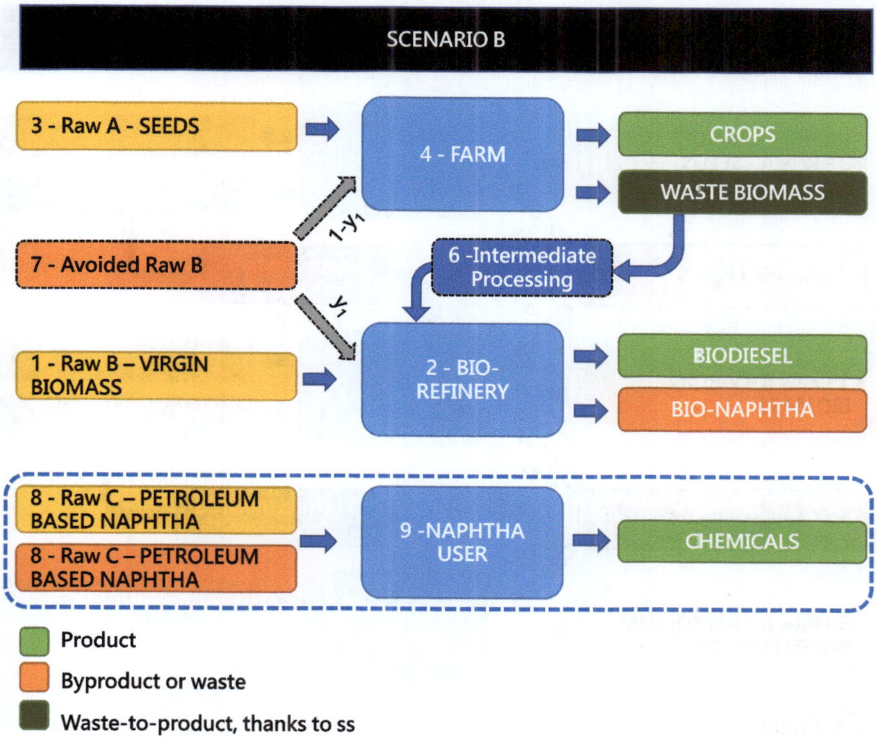

Fig. 4.5 Scenario B, two systems out of three are symbiotic

seeds and the chemical industry with the naphtha. There are no direct advantages; the allocations in the case of a multifunctional process can be made on the base of the products economic cost, both incoming and outgoing. In parallel, three separate systems coexist in the same industrial district.

- *Scenario B* (Fig. 4.5): the farm and the bio-refinery systems realize an industrial symbiosis; some advantage is achieved for the avoided production of raw material. The bio-refinery is partly fed (f(1 − x)) from virgin biomass and partly (f(x)) from waste biomass (treated as a waste product in scenario A). The 'advantages allocation rate' is represented in the form of the variable Y_1 to be discussed, which represents the fraction of who receives the benefits (50–50, 100–0, 0–100). The third system functions separately.
- *Scenario 3* (Fig. 4.6): the three activities—farm, bio-refinery and production of chemical products—achieve an industrial symbiosis, gaining advantages for avoided productions at the district level. Comparing to scenario B, more advantages are obtained for avoided production of virgin materials: while the bio-refinery was partly fed from virgin biomass (f(1 − x)) and partly (f(x)) from waste biomass, in the case of scenario C (extended symbiotic system) the

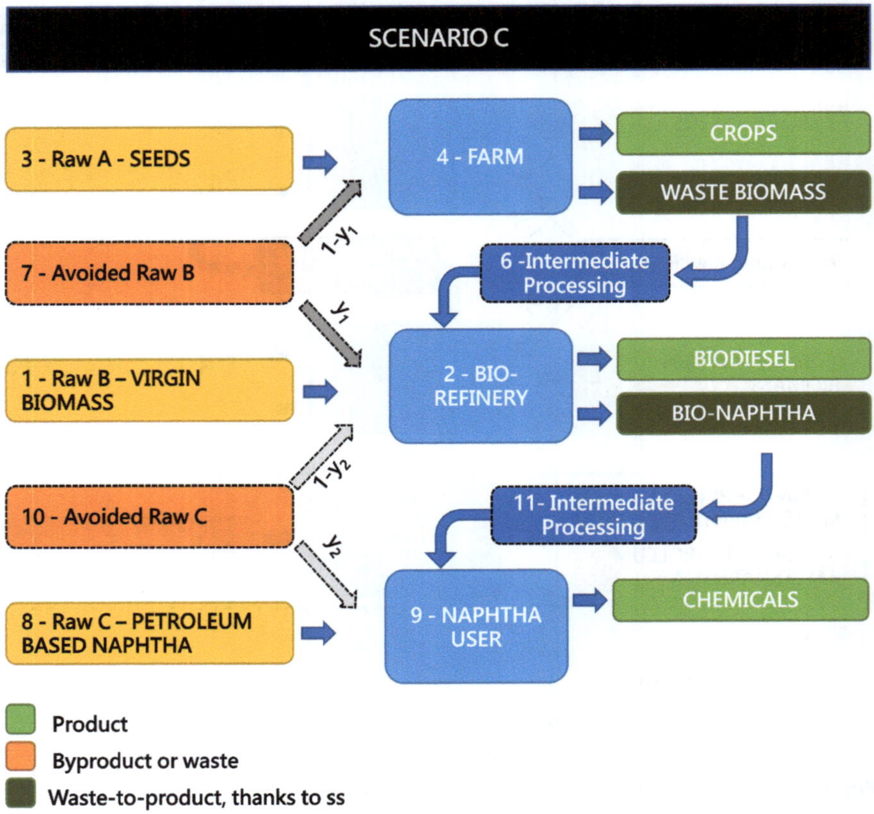

Fig. 4.6 Scenario C, three players symbiotic symbiosis

production of chemicals that uses naphtha, replaces the possible amount of naphtha of fossil origin with the bio-naphtha (the by-product exiting the bio-refinery). Furthermore, for the purpose of the analysis, it is assumed that the naphtha user process unit supplies only chemicals (no by-products), or the rest is considered as cut-off.

Main assumptions considered in the modelling phase are:

- UP2: The bio-refinery process yield gives from 8.83 kg of biomass (for example wheat straw) 1 kg of biodiesel and 0.37 kg of bio naphtha as a by-product (source: BTL process); this entails a mass allocation factor of 0.73 for biodiesel and of 0.27 for bio naphtha;
- UP4: To produce 1 kg of wheat grain (instance for crop), 0.032 kg of seeds are needed; the corresponding outgoing waste biomass;

- UP6 and UP11: The intermediate processing are for simplification purposes considered as transports: the bio-refinery has been considered 100 km far from the farm, where the waste biomass is produced; the ethylene production industry has been hypnotized 150 km far from the bio-refinery;
- $(1 - Y_1)$, Y_1, $(1 - Y_2)$ and Y_2 are the variable to be discussed, respectively, related to the avoided production of virgin biomass (UP1) and to the avoided production of petroleum-based naphtha (UP8), which represent the fractions of benefits to be allocated, according to predefined rules;
- During the modelling phase, the distribution of credits from the avoidance of Raw B and Raw C, when coexisting both (Scenario C), is assumed to be the same (the presented examples are both 50–50, or both 100–0, or both 0–100).

The symbiosis process could proceed linking separate systems, with the waste of a system becoming a power source, an input for another one. Furthermore, other different scenarios could be presented considering the by-product of an industry is used to feed—in different percentages—more symbiotic systems in the same context. The advantages should be than accurately and proportionally subdivided in all the interacting realities. Not changing the total impacts and benefit at a district scale, the costs and benefits in impact terms would be appreciable at a single system scale.

4.5.2 Model Output

The differences among the three systems output in the three scenarios are shown hereunder. Table 4.3 contains the formula used to assess the different scenarios impacts; Table 4.4 presents all the Unit Processes and their global warming potential emission factors (according to method GWP100 and IPCC). System 1 refers to the farm, system 2 to the bio-refinery, system 3 to the naphtha user.

For the transports, the following formula has been considered, according to the distances hypothesized (100 and 150 km, as reported before). Even significative changes in the distances, however, do not implicate relevant variations in the impacts.

$$I[UP6]_{100} = I[UP6]/10^3 \times c \times d_{100\,km} \quad (4.5)$$

$$I[UP11]_{150} = I[UP6]/10^3 \times d \times d_{150\,km} \quad (4.6)$$

$$I[UP7] = -I[UP1] \quad (4.7)$$

$$I[UP10] = -I[UP8] \quad (4.8)$$

Tables 4.5, 4.6 and 4.7 show the results obtained applying the formula presented in Table 4.2. In particular, the results emerging applying different benefits criteria are exposed.

Table 4.3 Scenarios and systems ruling formula

System	
Scenario A	
1	I [UP3] + I [UP4] + I [UP5]
2	*a**I [UP1] + I [UP2]
3	*b**I [UP8] + I [UP9]
1 + 2 + 3	I [UP3] + I [UP4] + I [UP5] +*a**I [UP1] + I [UP2] + *b**I [UP8] + I [UP9]
1 + 2	I [UP3] + I [UP4] + I [UP5] + *a**I [UP1] + I [UP2]
Scenario B	
1	I [UP3] + I [UP4]−y_1 *(c)* I [UP1]
2	(*a*−*c*) * I [UP1] + I [UP2] +I $[UP6]_{100}$ − (1−y_1) *(*c*)* I [UP1]
3	*b**I [UP8] + I [UP9]
1 + 2 + 3	I [UP3] + I [UP4]−y_1 * (*c*) * I [UP1] + (*a*−*c*) * I [UP1] + I [UP2] +I $[UP6]_{100}$−(1−y_1) * (*c*)* I [UP1] + *b**I [UP8] + I [UP9]
1 + 2	I [UP3] + I [UP4]−y_1 * (*c*) * I [UP1] + (*a*−*c*) * I [UP1] + I [UP2] +I $[UP6]_{100}$−(1−y_1) * (*c*)* I [UP1]
Scenario C	
1	I [UP3] + I [UP4]−y_1 *(c)* I [UP1]
2	(*a*−*c*) * I [UP1] + I [UP2] +I $[UP6]_{100}$−(1−y_1) * (x)* I [UP1]−(b−d) * (y_2) *I [UP8]
3	(b−d) *I [UP8] + I [UP9] + I $[UP11]_{150}$−d*(1−y_2) *I [UP8]
1 + 2 + 3	I [UP3] + I [UP4]−y_1 * (c) * I [UP1] + (a−c)* I [UP1] + I [UP2] +I $[UP6]_{100}$−(1 −y_1)*(c)* I [UP1]−(b−d)*(y_2)*I [UP8] +(b−d) *I [UP8] + I [UP9] + I $[UP11]_{150}$−d*(1−y_2)*I [UP8]
1 + 2	I [UP3] + I [UP4]−y_1 * (c) * I [UP1] + (a−c) * I [UP1] + I [UP2] +I $[UP6]_{100}$−(1−y_1) *(c)* I [UP1]−(b−d) * (y_2) * I [UP8]

With a, b, c and d as parameters relative to mass allocation for the production of the following products
a = 8.83 mass allocation for biodiesel production
b = 0.92 mass allocation for ethylene production
c = 0.609 mass allocation for wheat straw production
d = 0.37 mass allocation for bio naphtha production

Results and Discussion

Observing the output of the impact assessment and of the comparison about the global warming impacts of the systems, considering them separately and increasing the symbiotic degree, the consequent considerations are the following:

1. In a district where the industrial symbiosis is realized, globally 'everyone wins' in terms of impacts because global impacts are lowered at the district level;
2. In reality, this reduction depends on the intermediate process weight: if the intermediate processes are extremely onerous, the symbiosis could also become inconvenient;

Table 4.4 Process units GWP100 emission factors (Method: CML—IPCC)

Unit of process		Reference flow		Emission factor	
n.	Description			GWP100 (CML-IPCC)	
UP1	Biomass	1 kg	Wheat straw	0.046	kg CO_2
UP2	Bio-refinery	1 kg	Biodiesel (from 8.83 kg biomass)	6.463	kg CO_2
UP3	Seeds	1 kg	Seeds	0.017	kg CO_2
UP4	Farm	1 kg	Wheat grain (from 0.032 kg seeds)	0.484	kg CO_2
UP5	Waste biomass	1 kg	Biowaste	0.207	kg CO_2
UP6	Intermediate processing	1 ton km	Transported straw	0.083	kg CO_2
UP7	Avoided virgin biomass	−1 kg	Wheat straw	−0.046	kg CO_2
UP8	Petroleum-based naphtha	1 kg	Naphtha	0.332	kg CO_2
UP9	Naphtha user—ethilene	1 kg	Ethilene (from 0.92 kg naphtha}	1.450	kg CO_2
UP10	Avoided petroleum-based naphtha	−1 kg	Naphtha	−0.332	kg CO_2
UP11	Intermediate processing	1 ton km	Transported naphtha	0.083	kg CO_2

Table 4.5 Impact assessment results (application of formula of Table 4.2)—case with benefits shared 50–50

	Case 50–50					
	Scenarios	A kg CO_2	B kg CO_2	$\Delta_{B–A}$%	C kg CO_2	$\Delta_{C–A}$ %
Systems	**1**	0.71	0.49	*−31.22*	0.49	*−31.22*
	2	6.87	6.83	*−0.54*	6.74	*−1.87*
	3	1.76	1.76	*0.00*	1.58	*−10.23*
	1 + 2 + 3	9.33	9.07	*−2.76*	8.80	*−5.67*
	1 + 2	7.58	7.32	*−3.40*	7.23	*−4.61*

3. At the district level there is no interest nor difference in understanding/exploring the 'who gets the benefits' criteria, as there is no special beneficiary, but the whole district earns (see point 1);
4. If instead, the goal is to calculate the LCA of the different products and subjects of the district, there is a need to establish a policy of sharing the benefits between the symbiotic systems (the above mentioned 'who gets the benefits' criteria, represented from the **y** factor);

Table 4.6 Impact assessment results (application of formula of Table 4.2)—case with benefits all for the industry producing the material (100–0)

	Scenario 100–0					
	Scenarios	A kg CO_2	B kg CO_2	Δ_{B-A} %	C kg CO_2	Δ_{C-A} %
Systems	**1**	0.71	0.47	*−33.20*	0.47	*−33.20*
	2	6.87	6.85	*−0.33*	6.66	*−2.99*
	3	1.76	1.76	*0.00*	1.64	*−6.74*
	1 + 2 + 3	9.33	9.07	*−2.76*	8.77	*−5.99*
	1 + 2	7.58	7.32	*−3.40*	7.14	*−5.81*

Table 4.7 Impact assessment results (application of formula of Table 4.2)—case with benefits all for the symbiotic subject using the waste as input (0–100)

	Case 0–100					
	Scenarios	A kg CO_2	B kg CO_2	Δ_{B-A} %	C kg CO_2	Δ_{C-A} %
Systems	**1**	0.71	0.50	*−29.24*	0.50	*−29.24*
	2	6.87	6.82	*−0.74*	6.82	*−0.74*
	3	1.76	1.76	*0.00*	1.51	*−13.73*
	1 + 2 + 3	9.33	9.07	*−2.76*	8.83	*−5.35*
	1 + 2	7.58	7.32	*−3.40*	7.32	*−3.40*

5. At this point the value of the **y** factor can be discussed; depending on the situation and on the specific type of industry cooperating, there will be a different approach, in which the practitioner's experience, common sense and wisdom, in addition to other case-specific needs will play a fundamental role.
6. There seems to be the need to create standard rules to give value to **y** (50–50, 0–100, 100–0,...) that should be properly defined, at least for every analysis.

According to the results of the examined case, the advantage in the symbiotic system derives from the avoided impacts. Symbiotic processes allow cutting specific systems impacts such as waste treatment, production of virgin biomass and raw material C (naphtha) for chemical production. On the other hand, the additional impacts regard the two intermediate treatments (normally slightly more relevant than a transport), so although it is necessary to evaluate each case separately according to the type of process unit, the balance should be advantageous for the symbiotic system.

As general results following outcomes can be provided:

- In general terms, when a district performs an industrial symbiosis, single players change linear production of chemicals to circular production where the industrial symbiosis is realized. In such scenarios, additional impacts due to overconsumption and waste are avoided and environmental profiles tend to be lower, in presence of the same market conditions, at the district level. In such view, at district level there is no special beneficiary while the benefit is a cooperative result.

- More in detail, IS benefits at a single level depend on the intermediate additional process: in case the intermediate processes are extremely onerous, the symbiosis can become inconvenient.

4.6 Analysis Roadmap

Procedure suggestion for practitioners facing the analysis of the symbiotic process:

1. Start dividing the district into ***n*** isolated and closed subsystems
2. Performing LCA of ***n*** isolated systems (Fig. 4.7):
3. Identifying reciprocal replacements (Fig. 4.8):
4. Connecting two symbiotic systems and progressively adding the next ones connected, one at a time (Fig. 4.9).

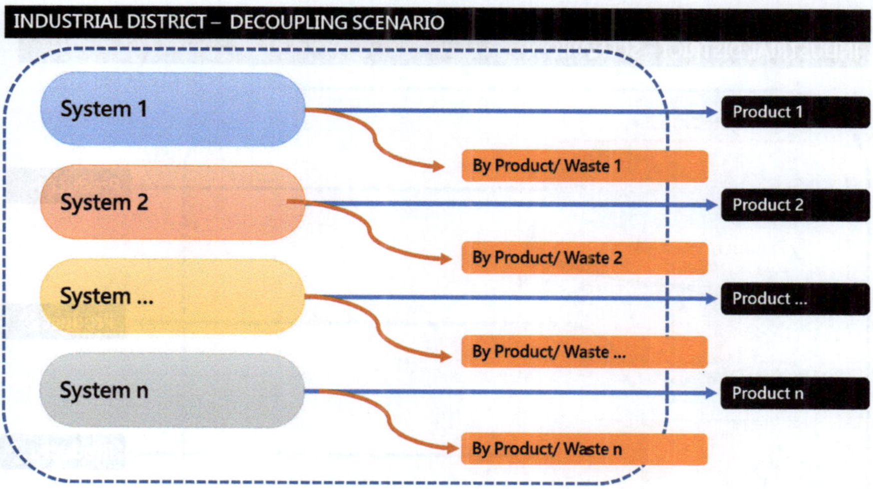

Fig. 4.7 LCA of *n* isolated systems scheme

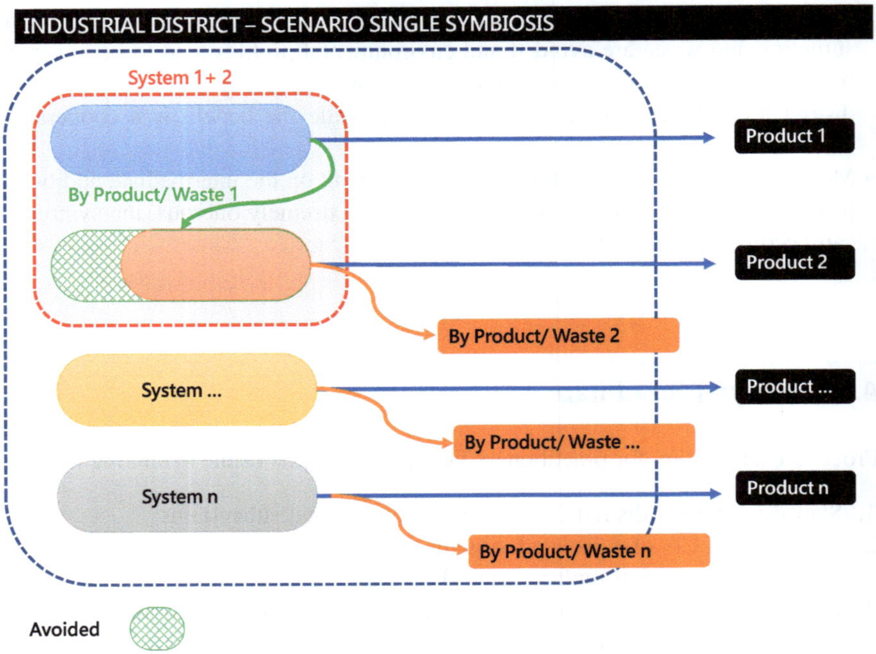

Fig. 4.8 Systems reciprocal replacements identification

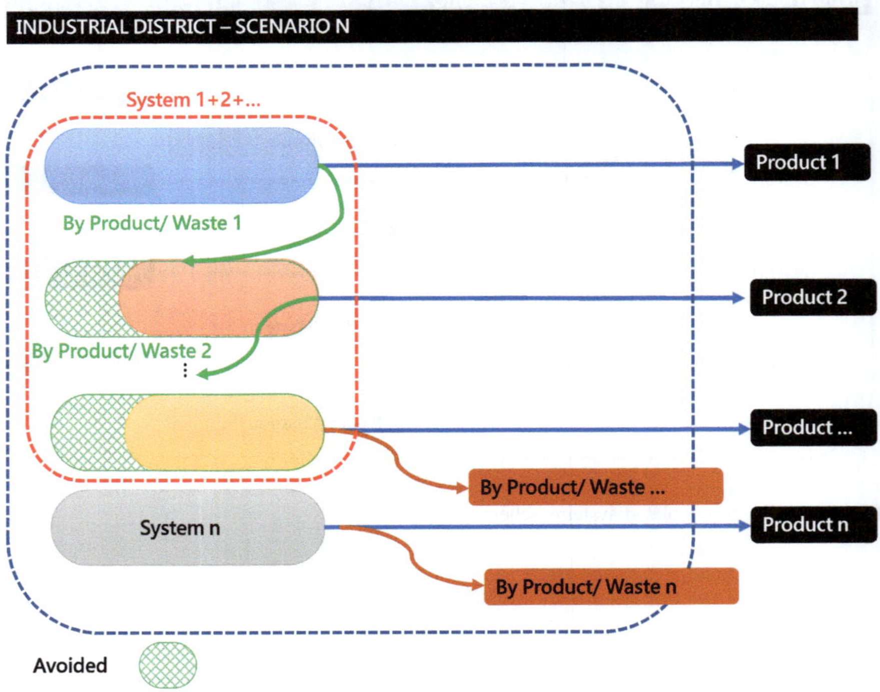

Fig. 4.9 Progressive connection of symbiotic systems

References

Ammenberg J, Baas L, Eklund METAL (2015) Improving the CO_2 performance of cement, part III: the relevance of industrial symbiosis and how to measure its impact. J Clean Prod 98:145–155

Baldassarre B, Schepers M, Bocken et al (2019) N,Industrial Symbiosis: towards a design process for eco-industrial clusters by integrating Circular Economy and Industrial Ecology perspectives, J Clean Prod 216:446–460

Basu P (2006) Combustion and gasification in fluidized beds. CRC press

Blengini GA, Busto M, Fantoni M et al (2012) Eco-efficient waste glass recycling: Integrated waste management and green product development through LCA. Waste Manag 32(5):1000–1008

Bridgwater AV (2003) Renewable fuels and chemicals by thermal processing of biomass. Chem Eng J 91:87–102

Brondi C, Cornago S, Ballarino A et al (2018) Sustainability-based optimization criteria for industrial symbiosis: the symbioptima case, Procedia CIRP69:855–860

Chen C, Habert G, Bouzidi et al (2010) LCA allocation procedure used as an incitative method for waste recycling: an application to mineral additions in concrete. Resour Conserv Recycl 54 (12):1231–1240

Chiodini A, Bua L, Carnelli L et al (2017) Enhancements in biomass-to-Liquid processes: gasification aiming at high hydrogen/carbon monoxide ratios for direct Fischer–Tropsch synthesis applications. Biomass Bioener 106:104–114

Curran T, Williams ID (2012) A zero waste vision for industrial networks in Europe. J Hazard Mater 207–208:3–7

Daddi T, Nucci B, Iraldo F (2017) Using Life Cycle Assessment (LCA) to measure the environmental benefits of industrial symbiosis in an industrial cluster of SMEs. J Clean Prod 147:157–164

De Souza-Santos ML (1989) Comprehensive modelling and simulation of fluidized bed boilers and gasifiers. Fuel 68:1507–1521

Dong H, Ohnishi S, Fujita T et al (2014) Achieving carbon emission reduction through industrial & urban symbiosis: a case of Kawasaki. Energy 64:277–286

Dong L, Liang H, Zhang L et al (2017) Highlighting regional eco-industrial development: life cycle benefits of an urban industrial symbiosis and implications in China. Ecol Model 361:164–176

Ekvall T, Finnveden G (2001) Allocation in ISO 14041—a critical review. J Clean Prod 9(3):197–208

European Parlament: Directive (EU) 2018/2001 of the European Parliament and of the Council of 11 December 2018 on the promotion of the use of energy from renewable sources (recast) (2018) (RED II) 328/82 Official Journal of the European

European Union—2018 update. Bioenergy policies and status of implementation. 2018. Country reports IEA Bioenergy

Huang M, Wang Z, Chen T (2019) Analysis on the theory and practice of industrial symbiosis based on bibliometrics and social network analysis. J Clean Prod 213:956–967

Jess A (1996) Catalytic upgrading of tarry fuel gases: a kinetic study with model components. Chem Eng Process 35:487–494

Ji P, Feng W, Chen B (2009) Production of ultrapure hydrogen from biomass gasification with air. Chem Eng Sci 64:582–592

Kerdlap P, Low JSC, Steidle R et al (2019) Collaboration platform for enabling industrial symbiosis: application of the industrial-symbiosis life cycle analysis engine. Proc CIRP 80:655–660

Kim HW, Ohnishi S, Fujii M et al (2017) Evaluation and allocation of greenhouse gas reductions in industrial symbiosis. J Industr Ecol 22(2):275–287

Lazarevic D, Martin M (2018) Life cycle assessment calculative practices in the Swedish biofuel sector: Governing biofuel sustainability by standards and numbers. Bus Strat Environ 27 (8):1558–1568

Liu Q, Jiang P, Zhao J et al (2011) Life cycle assessment of an industrial symbiosis based on energy recovery from dried sludge and used oil. J Clean Prod 19(15):1700–1708

Martin M, Svensson N, Eklund M (2015) Who gets the benefits? an approach for assessing the environmental performance of industrial symbiosis. J Clean Prod 263

Martin M, Wetterlund E, Hackl R, Holmgren KM, Peck P (2017) Assessing the aggregated environmental benefits from by-product and utility synergies in the Swedish biofuel industry. Biofuels

Martin M, Harris S (2018) Prospecting the sustainability implications of an emerging industrial symbiosis network. Resour Conserv Recycl 138:246–256

Mattila T, Helin T, Antikainen R, Soimakallio S, Pingoud K, Wessman H (2011) Land use in life cycle assessment. The Finnish Environment 24/2011

Mattila T, Lehtoranta S, Sokka L et al (2012) Methodological aspects of applying life cycle assessment to industrial symbioses. J Ind Ecol 16(1):51–60

Martin M (2015) Quantifying the environmental performance of an industrial symbiosis network of biofuel producers. J Clean Prod 102:202–212

Mohammed F, Biswas WK, Yao H et al (2016) Identification of an environmentally friendly symbiotic process for the reuse of industrial byproduct—an LCA perspective. J Clean Prod 112 (4):3376–3387

Mohammed F, Biswas WK, Yao H et al (2018) Sustainability assessment of symbiotic processes for the reuse of phosphogypsum. J Clean Prod 188:497–507

Perego C, Bianchi D, Bua L (2010) Biomass catalytic conversion to fuel and energy. La chimica e l'industria, 90–97

Pushkar S (2019) Modeling the substitution of natural materials with industrial byproducts in green roofs using life cycle assessments. J Clean Prod 227:652–661

Rank G (1972) The rectisol process for the selective removal of CO_2 and sulfur compounds from industrial gases. Chem Econ Eng Rev 4(5)

Røyne F, Hackl R, Ringström E, Berlin J (2018) Environmental evaluation of industry cluster strategies with a life cycle perspective: replacing fossil feedstock with forest-based feedstock and increasing thermal energy integration. J Ind Ecol 22(4):694–705.

Seto KE, Cameron JC, Daman K (2017) Panesar, Influence of fly ash allocation approaches on the life cycle assessment of cement-based materials. J Clean Prod 157:65–75

Sokka L, Lehtoranta S, Nissinen A, Melanen M (2011) Analyzing the environmental benefits of industrial symbiosis: life cycle assessment applied to a finnish forest industry complex. J Ind Ecol 15(1):137–155

Tijimensen MJA, Faaij APC, Hamelinck CN et al (2002) Exploration of the possibilities for production of Fischer Tropsch liqueids and power via biomass gasification. Biomass Bioen 23:129–152

Zampori L, Pant R (2019) Suggestions for updating the Product Environmental Footprint (PEF) method, EUR 29682 EN. Publications Office of the European Union, Luxembourg. ISBN 978-92-76-00654-1

Zennaro R, Ricci M, Bua L et al (2013) Syngas: the basis of Fischer-Tropsch. In: Maitlis PM, de Klerk A (eds) Greener Fischer-Tropsch processes for fuels and feedstocks. Wiley-VCH Verlag GmbH & Co. KGaA, pp 17–51

Part II
Application of LCA in Chemicals Process Modelling

LCA Application to Chemical Synthesis at Laboratory Scale

5

Martina Pini, Roberto Rosa, Paolo Neri and Anna Maria Ferrari

Abstract

Although a huge number of Green Chemistry metrics are available and more simply applicable with respect to LCA, their intrinsic less comprehensive nature could in some cases represent a limiting factor for a trustworthy evaluation of the environmental and human health impacts assessment associated to that specific chemical. Therefore, all of the chemical processes not specifically based on a chemical reaction (e.g., the extraction of a particular phytochemical compound from plant matrices) are far from being assessed by most of these metrics. LCA methodology-based evaluations for chemical processes are not limited by a mere chemical reaction equation, being also able to account for time and energy contributions together with all of the possible environmental loads associated to a particular process or product. This chapter will first overview the most widely employed Green Chemistry metrics. The possibility to integrate those metrics with the all-encompassing LCA methodology will be also accurately and critically discussed. This chapter will also furnish important recommendations and guidelines on when and at which extent the application of LCA should be highly suggested at a laboratory scale.

M. Pini · R. Rosa · P. Neri · A. M. Ferrari (✉)
Department of Science and Methods for Engineering,
University of Modena and Reggio Emilia, Modena, Italy
e-mail: annamaria.ferrari@unimore.it

S. Maranghi and C. Brondi (eds.), *Life Cycle Assessment in the Chemical Product Chain*, https://doi.org/10.1007/978-3-030-34424-5_5

5.1 Introduction

As previously reported in Chap. 3, Green Chemistry is based on the famous 12 principles proposed in the late 90s (Anastas and Warner 1998). The need to quantitatively and objectively assess the way a particular chemical synthesis meets the requirements highlighted in those principles led to the rapid development of a great number of different Green Chemistry metrics. These metrics rapidly started being used to support the conventionally applied ones (e.g., reaction yield, reaction time, conversion, and selectivity) in order for the most of the chemical and pharmaceutical industries to face the always more and more sustainable development requirements which need to be pursued as a consequence of the worldwide environmental politics.

Particularly, according to which of the 12 principles of Green Chemistry they refer to, Green Chemistry metrics can be classified into the following four different categories: mass metrics, environmental sustainability metrics, safety metrics, and energy metrics (Sheldon 2018). A schematic overview showing the kind of metric needed to accomplish each of the 12 principles of Green Chemistry is reported in Table 5.1.

From Table 5.1, it immediately appears clear that the assessment of how much a specific synthetic or preparation procedure of a target chemical compound or product meets the Green Chemistry requirements is an extremely complex issue. Indeed, it requires the accurate quantification of several parameters that, on their own, need different kinds of metrics to be quantified. Moreover, when the sustainability evaluation needs to be applied, its holistic nature makes any assessment even more complicated.

Table 5.1 Category of Green Chemistry metrics needed to accomplish selected Green Chemistry principles

Green Chemistry principles		Category of Green Chemistry metrics
1	Waste prevention	Mass
2	Atom efficiency	Mass
3	Less hazardous chemicals	Environmental sustainability
4	Design of safer products	Environmental sustainability/safety
5	Less amounts and safer auxiliaries	Mass/environmental sustainability/safety
6	Energy efficiency	Energy
7	Renewable raw materials	Energy
8	Avoid derivatization	Mass
9	Catalyst instead of stoichiometric reagents	Mass
10	Products able to degrade	Environmental sustainability
11	Real-time monitoring for pollution prevention	Environmental sustainability/safety
12	Safer processes	Safety

Indeed, in order to assess the sustainability of whichever process or product (thus including also the chemical syntheses or the preparation of chemical products, in general, at a laboratory scale), further considerations related to the economic- and society-related issues must be incorporated (Sheldon 2018), in order to pursue a Life Cycle Thinking (LCT) development philosophy. In this perspective, the three dimensions of sustainability can be accounted by incorporating Life Cycle Assessment (LCA), Life Cycle Cost (LCC), and Social Life Cycle Assessment (S-LCA) methodologies, thus leading to what is known as Life Cycle Sustainability Assessment (LCSA) framework (Ferrari et al. 2019).

5.1.1 Overview of the Existing Green Chemistry Metrics

Mass-based Green Chemistry metrics are mainly devoted to the quantification of generated waste, used reagents, and incorporation of starting atoms into products. Some of them have been briefly described in Chap. 3; here, they will be fully treated. Most of the proposed mass-based Green Chemistry metrics, whose definitions and mathematical expressions have been adapted from Ribeiro and Machado (2013), Andraos (2016), are reported in Table 5.2. Among them, the environmental factor, E_f, that recently turned 25 (Sheldon 1992, 2007), probably

Table 5.2 Main mass-based Green Chemistry metrics

Name (acronym)	Mathematical expression
Environmental factor (E_f)	$E_f = \frac{\text{total waste (kg)}}{\text{mass of product (kg)}}$
Mass intensity (MI)	$\text{MI} = \frac{\text{total mass used in a process (kg)}}{\text{mass of product (kg)}}$
Mass productivity (MP)	$\text{MP} = \frac{1}{\text{MI}} \cdot 100$
Atom economy (AE) for the generic reaction A + B → C	$\text{AE} = \left(\frac{\text{MW of } C}{\text{MW of } A + \text{MW of } B}\right) \cdot 100$
Atom utilization (AU)	$\text{AU} = \left(\frac{\text{mass of the product}}{\text{total mass of all the substances produced}}\right) \cdot 100$
Relative mass efficiency	$\text{RME} = \left(\frac{\text{mass of the product}}{\text{total mass of the stoichiometric reagents}}\right) \cdot 100$
Element (X) efficiency	$\text{XEE} = \frac{\text{mass of the element X in the product}}{\text{total mass of the element X in the stoichiometric reagents}}$
Carbon efficiency (CE)	$\text{CE} = \left(\frac{\text{moles of carbon in the product}}{\text{moles of carbon in the reactants}}\right) \cdot 100$
Wastewater intensity (WWI)	$\text{WWI} = \frac{\text{mass of process water (kg)}}{\text{mass of product (kg)}}$
Solvent intensity (SI)	$\text{SI} = \frac{\text{mass of solvents (kg)}}{\text{mass of product (kg)}}$
Effective mass yield (EMY)	$\text{EMY} = \frac{\text{mass of product (kg)}}{\text{mass of non-benign materials used (kg)}}$

represents the most widely employed one, due to the higher attention typically given to the outward materials' flux during a particular chemical process. However, more recently, Process Mass Intensity or simply Mass Intensity (MI) has been recognized as the best metric for parametrization of the overall materials' efficiency performance, also because it encompasses, in its mathematical expression, further metrics like atom economy (AE), reaction yield (*Y*), and relative mass efficiency (Andraos 2016).

Limitations of mass metrics have been highlighted by Ribeiro and Machado (2013). Above all, most of them have shown to be strongly dependent upon the characteristics of the reaction itself, such as stoichiometry, reaction yield, excess of one stoichiometric reagent, molecular weights of the reagents, and mass of auxiliary materials used. Therefore, their use for comparative purposes between different kinds of synthetic approaches toward a target molecule, compound, or product has been highlighted to possess limitations, thus requiring to be made with extreme caution (Ribeiro and Machado 2013).

More importantly, all of the proposed mass-based Green Chemistry metrics suffer from the major drawback of treating all of the chemicals employed in the synthesis or in the workup procedure on the same level concerning the environmental and human health damage or their hazardous characteristics. Actually, the Effective Mass Yield (EMY, reported in Table 5.2) is the only mass metric that tentatively accounts for the concept of hazardous applied to the reactants of a particular reaction (Hudlicky et al. 1999), so that sometimes it has also been classified among the environmental metrics (Albini and Protti 2016). However, the definition of benign or hazardous remained pretty subjective until the use of

Table 5.3 Main environmental and safety metrics

Name (acronym)	Mathematical expression
Environmental quotient (EQ)	$\mathrm{EQ} = E - \text{factor}\ (E) \cdot \text{Unfriendliness Quotient}\ (Q)$ *
Environmental impact for the inward materials' flux ($\mathrm{EI_{in}}$)	$\mathrm{EI_{in}} = \frac{\text{substrate}\ (g) \cdot Q_{tot_{substrate}} + solvent(g) \cdot Q_{tot_{solvent}}}{\text{product}\ (g) \cdot Q_{tot_{product}}} + \frac{\text{auxiliary materials}\ (g) \cdot Q_{tot_{auxiliary\ materials}} + \text{catalyst}\ (g) + Q_{tot_{catalyst}} + \cdots\cdots\cdots}{\text{product}\ (g) \cdot Q_{tot_{product}}}$ **
Environmental impact for the outward materials' flux ($\mathrm{EI_{out}}$)	$\mathrm{EI_{out}} = \frac{\text{waste}\ (g) \cdot Q_{tot_{waste}}}{\text{product}\ (g) \cdot Q_{tot_{product}}}$ **

*where the unfriendliness quotient (Q) was considered for example 1 for Nacl and 100–1000 for heavy metal salts (according to toxicity) [10]

**$[\mathrm{Qtot}]_j = \left(\sum_{i=1}^{i=n} Q_i\right)/n$ represents the total weighting factor for the jth category of substances among all of those used in the process (e.g. substrate, solvent, catalyst, auxiliary materials, etc.). It represents the mean value calculated among the i weighting factors each of which calculated for the ith impact category among the n considered.

environmental quotient (EQ, Table 5.3) metric, introduced by Sheldon (1994) as the E-factor multiplied by the so-called unfriendliness quotient Q, that started tentatively accounting for the environmental and human health effects related to a particular chemical in a more objective manner.

Exactly in this framework, Eissen and Metzger (2002) developed the software EATOS (Environmental Assessment Tool for Organic Syntheses), extending the concept of unfriendliness quotient to all of the substances employed or produced, thus introducing two environmental and safety impact metrics, EI_{in} and EI_{out} (Table 5.3), accounting for both inward and outward material fluxes, respectively. Particularly, by the introduction of environmental and safety information like, for example, LD50 and LC50 values, occupational exposure limit, octanol–water partition coefficient as well as H statements (even if yet in the form of R-phrases, thus accounting for safety criteria like flammability, explosiveness, corrosiveness, and more), those metrics allowed considering several environmental-, human health-, and safety-related impact categories. In detail, the impact categories the EATOS software allows considering are claiming of resources, risk, human toxicity, chronic toxicity, ecotoxicology, ozone creation, air pollution, accumulation, degradability, greenhouse effect, ozone depletion, nitrification, and acidification. More details on this latter tool and the algorithm at its basis will be given in the following Sect. 5.1.2.

None of the up to now mentioned metrics, however, allows accounting for the energy consumptions accompanying a particular chemical reaction or process. Indeed, even when moving from a laboratory scale to an industrial one, every energy-related consideration will be necessarily reformulated, some energy metrics like the energy efficiency as well as its opposite, i.e., the energy expenditure, associated to synthesis itself, to the workup procedure or to both, should be calculated also at a laboratory scale, by using the expressions reported in Table 5.4, in order at least to increase the awareness of the laboratory personnel on energy-related issues.

An example of the importance of the energy considerations at a laboratory scale can be found in the NOP (Nachhaltigkeit Organische Chemie Praktikum, i.e., Sustainability in the organic chemistry lab course) website (NOP website: https://www.oc-praktikum.de/nop/en-entry), where the conversion of carbonyl groups into acetals for protective purposes has been investigated in terms of energy

Table 5.4 Main energy metrics

Name (acronym)	Mathematical expression
Energy efficiency (E_E)	$E_E = \frac{\text{mass of product (kg)}}{\text{energy consumption (kJ)}}$
Specific productivity (sP)	$\text{sP} = \frac{\text{amount of product (mol)}}{\text{electric work (kWh)}}$
Energetic reaction expenditure (A_r)	$A_r = \frac{\text{energy consumption during reaction (W)}}{\text{mass of product (kg)}}$
Energetic workup expenditure (A_w)	$A_w = \frac{\text{energy consumption during workup (W)}}{\text{mass of product (kg)}}$
Energetic process expenditure (A_p)	$A_p = A_r + A_w = \frac{1}{E_E}$

consumptions related to the use of two conventional lab-scale heating apparatuses (namely, an oil bath and a heating mantle), in comparison with the use of microwave energy, highlighting the great advantages of this latter alternative energy, limitedly, of course, to the studied scenario. This was demonstrated to be due to the inevitable energy losses to components of the reaction apparatus (e.g., the reactor walls, silicone oil, etc.) that actually would not require to be heated at all. On the opposite, the peculiar heating mechanism arising from microwaves–matter interaction leads to the generation of heat directly inside the reaction mixture (according, of course, to its dielectric properties), thus to a significant reduction in the abovementioned energy losses.

This latter example highlights the importance of investigating the possible use of alternative energy sources in conducting a specific chemical synthesis. The consciousness of the existing alternatives to conventional heating strategies needs to be strongly chased starting from the laboratory scale. In the absence of promising small-scale results, they could be neither considered at all when moving to a larger scale. The same considerations can be easily applied to further less conventional energy sources, like, for example, ultrasound energy (Leonelli and Mason 2010) as well as to mechanochemistry-based procedures (Do and Friscic 2017).

Moreover, even if, as mentioned above, any energy-related considerations will typically require to be re-evaluated when moving to larger scale, it needs to be reminded that some innovative approaches allow the application of conventional as well as innovative heating strategies, without scaling the volume of the experiments conducted in the laboratory, but, however, allowing the preparation of larger amounts of chemicals. This is actually the case of continuous flow synthesis that was recently identified as a greener alternative to batch synthesis in the preparation of both active pharmaceutical ingredients and inorganic engineered nanomaterials (Dallinger and Kappe 2017; Vaccaro et al. 2014; Caramazana et al. 2017, 2018). Indeed, this latter approach allows reaching an enhanced heat and mass transport, a better control on the residence time, a higher reproducibility, an easier scalability, shorter processing time, together with an intrinsic increased safety. Among the recognized advantages of flow chemistry, safety is probably considered one of the most important one. Particularly, increased safety arises from the reduction in reaction volumes, rather than from the use of less hazardous chemicals, thus minimizing severities in case of accidents.

5.1.2 Overview of the Existing Tools for Easy Calculation and Visualization of Green Chemistry Metrics

Several tools and algorithms that allow easy calculation of some of the abovementioned Green Chemistry metrics are available and are typically freely accessible from the corresponding websites. They were recently reviewed in Andraos (2016), Andraos et al. (2016). The most widely investigated and reported ones are summarized in Table 5.5, where the metrics (and their nature) each of them allows calculating are reported as well.

Table 5.5 List of the main tools available for the calculation of green metrics

Tool/algorithm	Metrics calculated (category of Green Chemistry metrics)[a]	Availability/cost	Visual aid
ACS PMI calculator	MI (m)	Free online calculator (ACS Green Chemistry Institute Pharmaceutical Roundtable 2018)	No
Andraos algorithm	Y (m), AE (m), RME (m), SF (m), MRP (m)	Spreadsheet available as supporting information in Andraos works (Andraos 2009)	Histograms and radial polygons
EcoScale	EcoScale score (m, s)	Free online calculator	No
Green star	GSAI (m, env, s, en)	Spreadsheet easily developing by following the criteria and scores proposed in Ribeiro et al. (2010)	Radial polygons
EATOS	MI (m), E_f (m), EI_{in} (env, s), EI_{out} (env, s)	Free of charge, downloadable at (Eissen and Metzger EATOS software website)	Histograms

where m = mass-based metric, env = environmental metric, s = safety metric, en = energy metric

Starting from tools that allow calculating only mass-based metrics, the American Chemical Society (ACS) Process Mass Intensity (PMI) calculator is a spreadsheet that can be found at the ACS website (2018) and allows directly calculating mass intensity metric for the whole chemical process (and indirectly the E-factor by the relation PMI = E_f + 1), accounting for both single reactions and linear or convergent synthetic plans.

Similar spreadsheets are at the basis of the Andraos algorithm (Andraos 2005). They are typically available as supporting information accompanying Andraos works (Andraos 2009). They allow considering and calculating reaction yield (*Y*), Atom Economy (AE), Relative Mass Efficiency (RME) and two further metrics, namely, Stoichiometric Factor (SF, (5.1)) and Material Recovery Parameter (MRP, (5.2)). These five mass-based Green Chemistry metrics can be then concurrently accounted by normalizing them into an overall Vector Magnitude Ratio (VMR) calculated by (5.3).

$$\mathrm{SF} = \frac{\text{actual mass of reagents (g)}}{\text{stoichiometric mass of reagents } (g)} \tag{5.1}$$

$$\mathrm{MRP} = \frac{\mathrm{SF} \cdot \text{mass of target product (g)}}{\mathrm{SF} \cdot \text{mass of target product (g)} + Y \cdot \mathrm{AE} \cdot \text{mass of auxilieries (g)}} \tag{5.2}$$

$$\mathrm{VMR} = \frac{1}{\sqrt{5}}\left[(\mathrm{AE})^2 + (Y)^2 + (\mathrm{RME})^2 + \left(\frac{1}{\mathrm{SF}}\right)^2 + (\mathrm{MRP})^2\right]^{\frac{1}{2}} \tag{5.3}$$

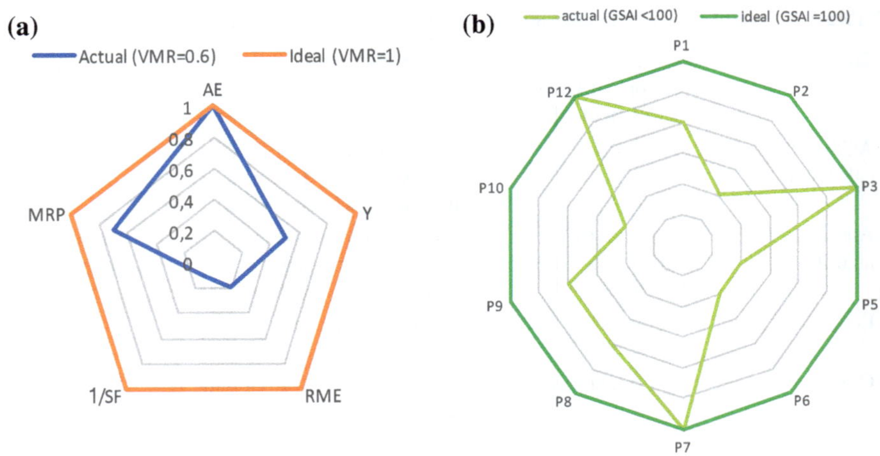

Fig. 5.1 **a** Example of radial polygons visualization aid for the Andraos algorithm applied to a generic synthesis with a VMR of 0.6. **b** Visual aid of Green Star approach, allowing calculating GSAI

The Andraos spreadsheets also allow visualizing the calculated metrics as diagrams represented by radial polygons (or even histograms) as those reported in Fig. 5.1a for a generic linear synthesis characterized by a VMR value of ca. 0.6.

EcoScale (Van Aken et al. 2006) is an online available calculator tool (EcoScale calculator website 2006) allowing determining penalty points to be subtracted to the ideal reaction value of 100, where the ideal reaction is defined as reaction in which "the substrate A undergoes a reaction with, or in the presence of, inexpensive compound B, to give product C in 100% yield at room temperature, with a minimal risk for the operator and a minimal impact on the environment" (Van Aken et al. 2006). The penalty points are arbitrary assigned according to reaction yield, price, safety (based on the hazard warning symbols limitedly to the reagents), experimental setup used, temperature/time necessary, and the workup procedure to be applied. Therefore, the higher the EcoScale value is the greener that particular reaction is considered. The obtained EcoScale value tentatively accounts for issues related to the mass of the substances employed, to the risk associated to them as well as to energy-related considerations (by considering the experimental apparatus as well as time and temperature employed in the reaction). However, this is done in an arbitrary way, and, above all, in a qualitative/semiquantitative way rather than giving quantitative data.

A similar approach is at the basis of the Green Star tool introduced in 2010 (Ribeiro et al. 2010) and more recently updated (Duarte et al. 2015). It determines merit points (ranging from 1 to 3, with 3 representing the greenest attribute) that are assigned to a particular linear chemical synthesis on the basis of its degree of accomplishment of 10 of the 12 principles of Green Chemistry (indeed, principles 4 and 11 are excluded). As an example, the way that specific synthesis accomplishes principle 1 of Green Chemistry is determined as follows:

- a merit score of 3 is assigned if the generated waste is innocuous;
- a merit score of 2 is assigned if the generated waste involves a moderate risk to human health and environment; and
- a merit score of 1 is assigned if the waste involves high risk to human health and environment.

The risk is accounted by considering the hazard symbols of the substances involved. The same approach is then applied to the other nine Green Chemistry principles of the 10 considered by this approach. A useful visual representation of results can be realized by using radial polygons (i.e., a decagon in this case) as the one reported in Fig. 5.1b. A Green Star Area Index (GSAI) is then calculated by the following (5.4), where p_i is the score value obtained by the actual synthesis considered (assuming values of 1, 2, or 3) in the accomplishment of the *i*th Green Chemistry principle.

$$\mathrm{GSAI}\,(\%) = 100 \cdot \left(\frac{\text{Actual area}}{\text{Ideal area}}\right) = \left[\frac{\sum_{i=1}^{i=10}(p_i p_{i+1})}{0.80}\right] \quad (5.4)$$

Although these latter two tools tentatively account also for time- and energy-related issues, together with environmental and safety ones, their main limitation is that they do not account for relative masses of input as well as waste materials, and they can be only applied to single syntheses and not to more complex synthetic plans.

The software EATOS has been already introduced in the previous paragraph. This software, together with its user manual, is freely downloadable after a simple registration procedure at the website of the inventors (Eissen and Metzger 2001; 2019), and its running only requires separate Java script program. Despite the calculation of mass-based Green Chemistry metrics, namely, E-factor and mass intensity (Table 5.2), it allows calculating the two environmental and safety impact metrics discussed above, i.e., EI_{in} and EI_{out} (Table 5.3), while any energy-related consideration is disregarded by this tool.

As reported in Table 5.3, EI_{in} and EI_{out} metrics account for the Environmental Impact (EI) of the inward (in) and outward (out) material fluxes, respectively, and this is realized by multiplying each employed or produced *j*th substance for its total weighting factor Q_{tot}, the latter representing the mean value calculated among the substance-specific weighting factors Q_i, where the index *i* accounts for the *i*th weighting category among those that can be considered by EATOS software, i.e., claiming of resources, risk, human toxicity, chronic toxicity, ecotoxicology, ozone creation, air pollution, accumulation, degradability, greenhouse effect, ozone depletion, nitrification, and acidification. Each Q_i assumes values comprised between 1 and 10, on the basis of arbitrary classification made by the internal algorithm itself, which depends on the specific characteristics of the substance considered. As an example, concerning the impact category claiming of resources, the corresponding $Q_{\text{claiming of resources}}$ is calculated by the following (5.5) (Eissen

Table 5.6 Weighting factor Q values for the impact category "claiming of resources" according to the price of a substance

$Q_{\text{claiming of resources}}$	Price range (€/g)	
	From	To
1	For free	0.10
2	0.10	0.24
3	0.24	0.56
4	0.56	1.34
5	1.34	3.17
6	3.17	7.51
7	7.51	17.80
8	17.80	42.22
9	42.22	100.13
10	100.13	∞

and Metzger EATOS user manual website), which leads to the classification of the substances, according to their price, as reported in Table 5.6.

$$Q_{Claiming\ Of\ Resources} = \left\lceil \frac{ln\left(\frac{Price\left(€/g\right)}{0.0178}\right)}{0.8635} \right\rceil \tag{5.5}$$

Similar calculations and classifications are obtained also for the further impact categories. The EATOS function weighting of weighting allows selecting the weighting categories to be considered in the study both for the input and for the output materials' flow. Particularly, as concerning the input materials, the EATOS software allows selecting only the weighting categories claiming of resources and risk, while in the case of the output materials, all the remaining weighting categories can be selected.

The visual aid of this tool is composed of evaluation histograms (one for each metric the software allows calculating) that can be used for comparative purposes of different synthetic strategies. In Fig. 5.2, the results related to four different syntheses of 4-methoxyacetophenone, as proposed by the case studies in the EATOS software database, are reported. Particularly, procedures a–d are those reported in the examples and are referred to well-known published procedures (Hünig et al. 1979, Becker 2001; Freese et al. 1999; Vogel 1978).

One limitation represented by a similar approach is the poor accessibility to the algorithm itself making most of the calculations hidden from the user, who only can understand the way the algorithm works by an accurate study of the software user guide. Further limitations have been recently highlighted by Cespi et al. (2016). However, the main drawback of this tool is the lack of any energy-related consideration.

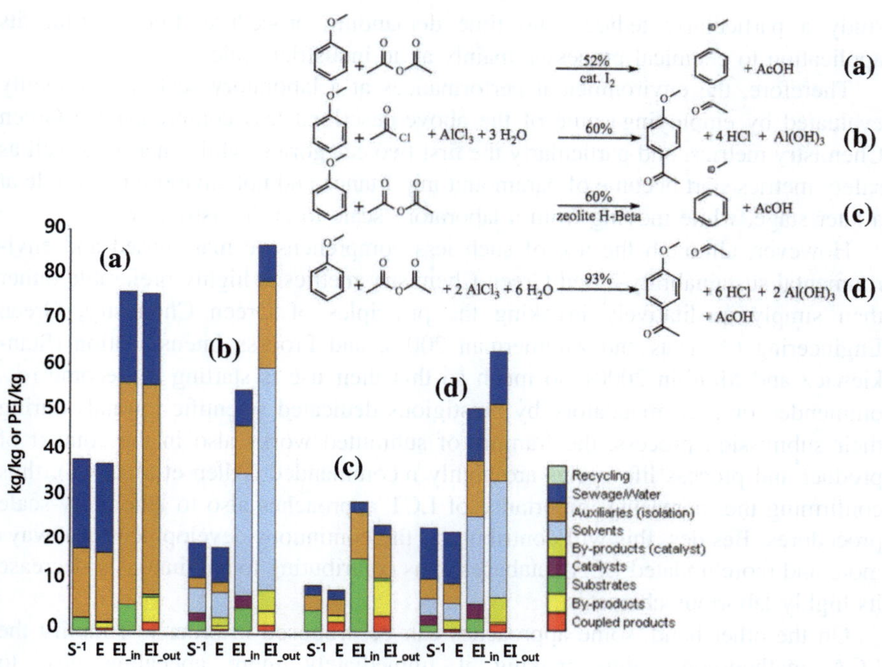

Fig. 5.2 Evaluation histograms for four different synthetic procedures (a–d) of 4-methoxyacetophenone, according to published procedures (Hünig et al. 1979; Becker 2001; Freese et al. 1999; Vogel 1978) respectively

Nevertheless, as it will be highlighted in the next section, EATOS software will be recommended as the tool of choice to be applied in synergy with more comprehensive life cycle assessments, in the evaluation of the greenness of chemical synthesis and preparation procedures at a laboratory scale.

5.2 Guidelines for the Application of LCA Methodology to the Laboratory-Scale Preparation of Chemicals: Why, When, and How?

This section aims at proposing some guidelines and related strategies for performing greenness evaluation of particular syntheses or preparation procedures, in general, for chemical compounds at a laboratory scale, by applying the LCA methodology.

Indeed, it is well known that the main limitation of LCA methodology is probably represented by the difficulty to source all the necessary inventory data, since most of the available databases are still limited in their content, even if under continuous updating. This contributes to make performing a life cycle assessment study a particularly tedious and time demanding procedure, thus limiting its application to chemical processes mainly at an industrial scale.

Therefore, the environmental performances at a laboratory scale are limitedly evaluated by employing some of the above-described less comprehensive Green Chemistry metrics, and particularly the first two categories, while energy as well as safety metrics start become of paramount importance and not anymore negligible at a later stage, while moving from a laboratory scale to an industrial one.

However, although the use of such less comprehensive mass-based and environmental sustainability-based Green Chemistry metrics is highly preferable rather than simply qualitatively invoking the principles of Green Chemistry, Green Engineering (Anastas and Zimmerman 2003), and Process Intensification (Stankiewicz and Moulijn 2000), so much so that their use is starting to become recommended or even mandatory by prestigious dedicated scientific journals during their submission process, the framing of submitted works also in the context of product and process life cycles are highly recommended (Allen et al. 2015), thus confirming the increasing importance of LCT approaches also to laboratory-scale procedures. Besides, this will contribute to the continuous development of always more and more updated LCA databases, thus contributing to continuously decrease its highly laborious character.

On the other hand, some approaches can be proposed in order to simplify the LCA methodology, thus making it immediately more appealing also to laboratory-scale applications. These latter simplifications include, for example, the replacement of some chemicals, which are absent from the LCA databases considered, with their more closely related analogue. A representative example is given by the environmental assessment of titanium dioxide photocatalysis versus conventional thermal conditions applied to the functionalization of heterocyclic compounds, performed by Ravelli et al. (2010). Indeed, in that work, the Life Cycle Inventory (LCI) step of the LCA study considered acetic acid instead of trifluoroacetic acid as well as hydrogen peroxide (since the only related redox-active material in the database) instead of hydroxylamine-O-sulfonic acid.

Similarly, the same research team later on, in performing the quantification of the environmental and human health burdens associated to the chemical synthesis of rose oxide (i.e., the mixture of *cis*- and *trans*-4-methyl-2-(2-methylprop-1-en-1-yl)tetrahydro-2*H*-pyran), considered a generic "organic chemical" during the LCI instead of Bengal rose (i.e., 4,5,6,7-tetrachloro-2′,4′,5′,7′-tetraiodofluorescein), due to the absence of close relationship with any of the compounds present in the considered databases (Ravelli et al. 2011). In the same work, all the aspects connected with the isolation of β-citronellol, like the growing of the vegetable, treating the agricultural wastes, the obtainment of the essential oil, and the subsequent extraction of the alcohol, were considered for the closest related cultivation found in the database, i.e., *sorghum* (Ravelli et al. 2011).

A particularly noteworthy finding of those research works was the close parallelism of the LCA results with those obtained by using the software EATOS, even considering the extremely different nature of the two approaches. Of course, when the amount of energy gives a major contribution, a partial loss of parallelism in the results obtained with the two approaches is expected.

Therefore, in alternative to the above-described simplified LCA approach in a chemical laboratory-scale scenario, where the choice between different synthetic strategies for the preparation of a target chemical compound needs to be pursued, one main recommendation would be to apply first the less comprehensive metric represented by EATOS tool, and the only subsequent application of LCA methodology to an already environmentally optimized chemical synthesis. Due to the saved computational efforts of this latter approach, the subsequently applied LCA study could in all likelihood be done without the previously made approximations. Indeed, the lack of inventory data can be overcome by accurate search in the scientific literature and patent databases with the aim to correctly model all of the processes and subprocesses leading to the preparation of the necessary chemical precursors absent in the considered databases. In this framework Pini et al. (2015), for example, created new processes starting from literature data and referred to precursors existing in the used database (i.e., Ecoinvent one) in order to model titanium tetrachloride and titanium tetraisopropoxide chemicals employed for the environmental assessment of the hydrolytic sol–gel synthesis of anatase TiO_2 nanoparticles. Similarly, Zerazion et al. (2016) used scientific literature data in order to model all of the background processes necessary for the obtainment of acetylacetone, tributyl borate, and vanillin, which were, on their own, precursors of curcumin molecule whose synthesis and extraction from natural resources were environmentally assessed, as it will be detailed in the next paragraph.

The here proposed recommendation concerning EATOS as the less comprehensive green metric/tool of choice is based on several further considerations besides the parallelism of its results with those of LCA. Particularly, in addition to being a free of charge software (Eissen and Metzger 2001; 2019), it allows easily calculating both mass-based metrics (including MI) and environmental- and safety-related metrics, as already reported in Table 5.5, that are of immediate interpretation due to its visual aid.

Moreover, an excellent parallelism of its results with those obtained by LCA methodology has also been demonstrated for the synthesis of inorganic materials (Pini et al. 2015), and this is a further fundamental finding, since the diffusion of green metrics parameters also to inorganic chemistry and materials science fields appears of paramount importance, bearing in mind the yet significant lack of information on long-term evaluation of environmental as well as human health impacts of engineered nanomaterials.

In order to partially overcome EATOS limitation of completely disregarding any energy contributions, the concurrent calculation of one or more energy metrics among those previously reported in Table 5.4 should be recommended when performing the starting evaluation (prior to the LCA one), at least if the comparative study comprises synthetic procedures at the basis of which there are significantly different energy contributions. Indeed, this is not the case, for example, of combinatorial chemistry studies, typically involving, a few small changes in the molecular structure of one or more precursors, and not significant changes in the conditions under which the reaction is conducted. Therefore, in combinatorial chemistry studies, the application of less comprehensive Green Chemistry metrics/tools, like the here recommended EATOS software, could be at a first stage the optimal compromise.

On the opposite, LCA methodology represents the sole feasible approach when the environmental and human health impacts need to be calculated and, above all, compared for preparation procedures of chemicals that fundamentally differ in their intrinsic nature. This is, for example, the case of a particular chemical synthesis of a target molecule in comparison with its extraction from natural resources. Indeed, the extraction of phytochemical compounds from plant matrices is typically considered a greener alternative with respect to the direct chemical synthesis, mainly when the concepts and principles of Green Analytical Chemistry (Armenta et al. 2008) and those of Green Extraction (Chemat et al. 2012) are pursued and applied. However, as it will be detailed in the next section, with a dedicated case study, these kinds of conclusions cannot be drawn in a trustworthy manner until a detailed comparison, performed with the LCA methodology, is performed.

5.3 Case Studies

In this section, some case studies will be presented in which the LCA methodology was applied to the laboratory-scale preparation of selected chemical compounds of particular relevance in different application fields, comprising the chemical synthesis of target organic molecules, as well as the direct extraction of phytochemical compounds from vegetable matrices, together with the inorganic synthesis of engineered oxide-based 0-D nanomaterials.

Moreover, the possibility of synergies of LCA methodology with less comprehensive metrics, particularly the mass-based and the environmental ones obtainable by employing the software EATOS, will be discussed, also on the basis of the considerations and recommendations made in the previous section.

The fundamental importance of the concurrent application of LCA methodology in the hereafter presented case studies will be highlighted and, above all, justified on the basis of the obtained results.

5.3.1 Chemical Synthesis of a Lactose Derivative to Be Employed in an Innovative Degreasing Formulation for the Tanning Cycle of the Leather Manufacturing

4-[6′-deoxy-1′-(2,3:5,6-di-*O*-isopropylidene-1,1-dimethoxyaldehydo-D-glucos-4-yl)-3′, 4′-*O*-(isopropylidene)-β-D-galactopyranos-6′-yl]piperazine hydrochloride (compound **5** Fig. 5.3) represents the fundamental ingredient of an innovative degreasing formulation (named EDF20) with an ethoxylated content reduced down to 25% (with respect to values of 45–55% typically used in most of the commercially available products) to be employed in the tannin cycle within leather manufacturing. Compound **5** and EDF20 degreasing formulation were, respectively, synthesized and optimized in the framework of a European LIFE+ project entitled "Environmentally-friendly natural products instead of chemical products in the degreasing phase of the tanning cycle" and whose acronym is Ecodefatting (Life+ Ecodefatting project website).

The scheme for the synthesis of compound **5** is reported in Fig. 5.3. It is a four-step synthetic pathway whose details can be found elsewhere (Rosa et al. 2017). Briefly, lactose is treated with dimethoxypropane (DMP) to achieve compound **2**. Subsequently, the primary alcoholic function undergoes a regioselective tosylation and then a nucleophilic substitution to replace the tosyl group of compound **3** by the piperazinyl one (Bianchini et al. 2014). In the last step, the as-obtained compound **4** is treated with stoichiometric hydrochloric acid solution in ethanol in order to deliver compound **5**.

3200 g of EDF20 degreasing formulation is then prepared by typically mixing at 20 °C an aqueous solution of 160 g of compound **5**, with 640 g of ethoxylated iso-decylalcohol 5mEO, 800 g of an ester compound, naturally derived, and 160 g of hexylene glycol, followed by dilution with water.

The need of performing an environmental and human health impacts assessment was mainly driven by the fact that leather manufacturing represents one of the most impacting industrial sectors, with alternatives to all of its subprocesses which are practically being published or proposed on a daily basis, without however, in the majority of the cases, presenting quantitative and trustworthy data on the environmental burdens associated to the newly proposed strategies/alternatives.

In this framework, EDF 20 poses itself as a valuable alternative product to be employed in the hide degreasing phase of the preparatory stage, which is of paramount importance in order to allow the action of the tanning agents during the subsequent tanning phase.

In order to assess all of the stages leading to the preparation of EDF 20, the use of LCA methodology instead of less comprehensive metrics results almost mandatory. Indeed, comprising a step (i.e., the fifth one) at the basis of which there

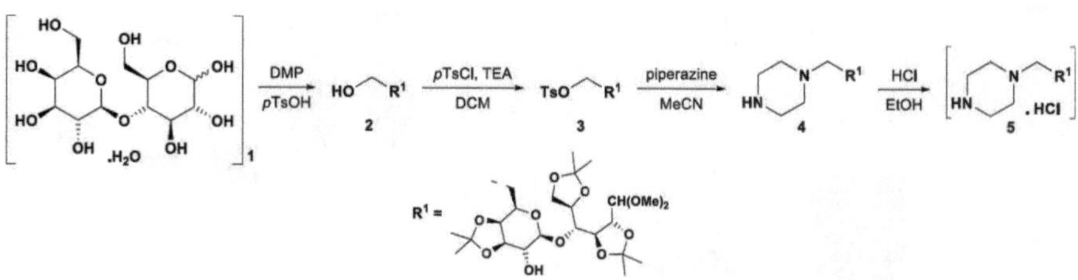

Fig. 5.3 Reaction scheme for compound **5**. Reproduced from Rosa et al. (2017) with permission from The Royal Society of Chemistry

is not a stoichiometric reaction, most of the Green Chemistry metrics and associated tools cannot be applied in order to operate a trustworthy comparison of the environmental performances characterizing the five different steps.

At a first approximation, however, the software EATOS was initially employed in order to perform an environmental evaluation limitedly to the mere chemical nature of substances employed in all the different stages of the four-step synthesis of compound 5. Although its use can help in focusing the future improvement efforts on the chemical characteristics of the substances employed (due also to the fact that very often the major contributions to the overall impact is revealed to be the energy contributions by LCA method, thus partially hiding the loads associated with the chemicals themselves), in the present case, the parallelism between the two methodologies was proven to be extremely poor, as easily deducible from the comparison of their obtained results as reported in Fig. 5.4. Indeed, according to EATOS results, the third step results the most impacting (possessing the highest values of both EI_{in} and EI_{out}), while a more comprehensive evaluation even highlights its positive impact on the overall evaluation (possessing negative *mPt* values). This latter aspect was due to the significant recovery of acetonitrile occurring during step 3, thus avoiding chemical oxygen demand in water. On the opposite by applying the LCA methodology, the first two steps result the most impacting ones, with the electric energy consumption during the complex workup procedures of those intermediates, significantly affecting the most impacting categories.

Therefore, as already advised in the few guidelines proposed in the previous section, it needs to be restated how the application of the sole EATOS software (or other less comprehensive Green Chemistry metric) to cases in which the energy contribution is significantly different among steps of a single procedure or among different procedures is highly discouraged, while the application of LCA methodology should be preferred.

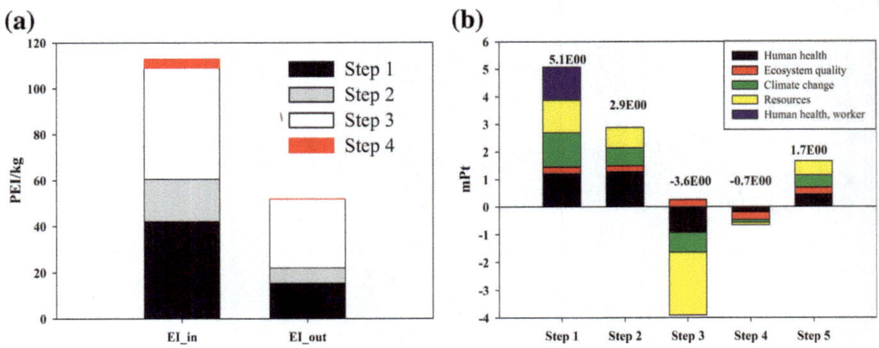

Fig. 5.4 **a** Contributions of the four synthetic steps to the EI_{in} and EI_{out} EATOS metrics. **b** Single-score evaluation LCA results for all the five steps leading to the laboratory-scale production of 3200 g of EDF 20. Reproduced from Rosa et al. (2017) with permission from The Royal Society of Chemistry

5.3.2 Direct Organic Synthesis Versus Extraction from Natural Matrix: The Case of Curcumin Molecule

Curcumin, i.e., [1,7-bis(4-hydroxy-3-methoxyphenyl)-1,6-heptadien-3,5-dione] (i.e., the product in Fig. 5.5) was selected as a target organic molecule to the synthesis of which the LCA methodology was applied, and the reason is twofold. First, curcumin is widely known for its excellent antioxidant, anti-inflammatory, and anticancer properties, so much so that it is actually under evaluation for approval by the U.S. Food and Drug Administration (FDA). Moreover, curcumin molecule can also be experimentally obtained, by its direct extraction from *Curcuma longa* L. dried rhizomes, with the opportune solvent and extraction procedure.

Therefore, it represents an excellent example for evaluating preparation procedures of different nature toward the obtainment of a targeted chemical compound (Zerazion et al. 2016).

Particularly, a synthetic pathway based on a modified Pabon reaction (Pabon 1964) was evaluated as developed by Ferrari et al. (2011). The reaction scheme considered is reported in Fig. 5.5.

Despite the chemical synthesis of curcumin, its direct extraction from Curcuma longa L. dried rhizomes was assessed, by considering both a conventional soxhlet-based extraction procedure and a microwave-assisted extraction, according to the experimental setup reported in Fig. 5.6 (Rosa et al. 2018).

The results of the study, expressed in terms of single-score evaluation, are summarized in Fig. 5.7.

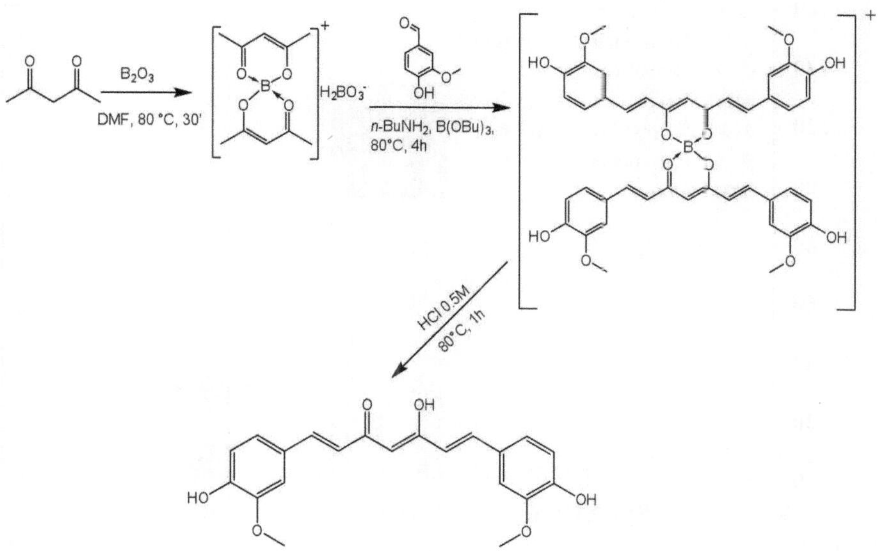

Fig. 5.5 Synthesis of curcumin according to the modified Pabon reaction as optimized by Ferrari et al. (2011). Reproduced from Zerazion et al. (2016) with permission from The Royal Society of Chemistry

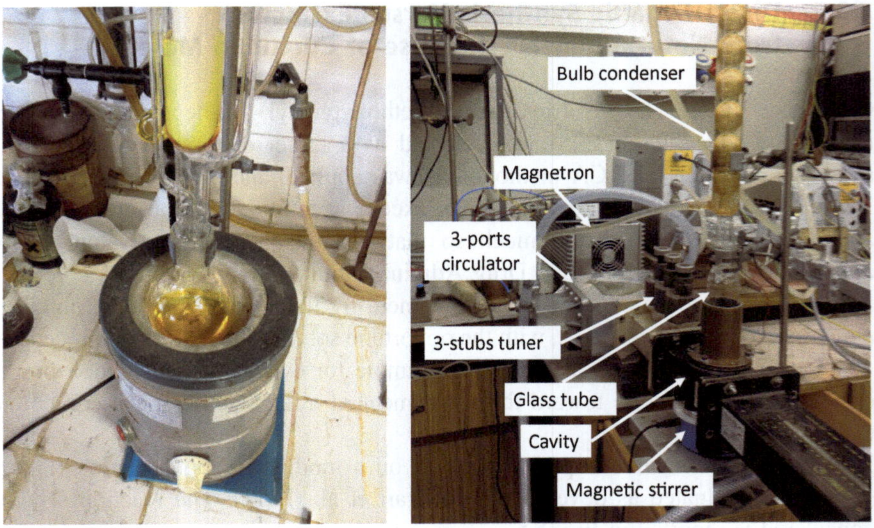

Fig. 5.6 Conventional Soxhlet-based apparatus (left) and microwave-assisted extraction applicator (right). Originally published in Rosa et al. (2018) under Creative Commons Attribution 3.0 Unported License. Available from https://doi.org/10.5772/intechopen.73651

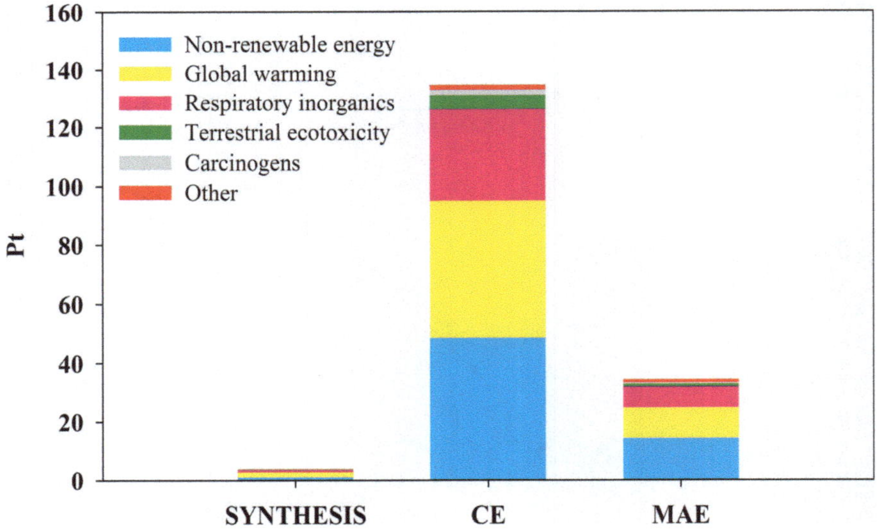

Fig. 5.7 Comparative analysis of curcumin life cycle produced by the three different techniques considered: synthesis, Conventional Soxhlet-based Extraction (CE), and Microwave-Assisted Extraction (MAE). Reproduced from Zerazion et al. (2016) with permission from The Royal Society of Chemistry

Rather surprisingly, the LCA methodology allowed highlighting that extraction from natural resources does not always can be considered a greener alternative to the direct chemical synthesis of a particular phytochemical compound. Indeed, in a chemical laboratory-scale scenario, the considered chemical synthesis of curcumin (even if not optimized from an environmental perspective, e.g., see the use of DMF as solvent from Fig. 5.5) can be considered less impacting with respect to its extraction from *Curcuma longa* L. (in the experimental conditions evaluated) and the main reason is surely to be attributed to the higher reaction yield (ca. 60%) with respect to those obtainable by extraction (ca. 6–7%). This work also contributed to prove the green characteristics of MAE with respect to conventional extraction procedures, at least at the considered laboratory scale.

This work represents a further example of the necessity of applying a more comprehensive approach (i.e., the LCA methodology), when the environmental evaluation is needed to be performed on chemical processes of completely different nature, and the results compared, in order to tentatively establish the greener option, for the preparation of a target chemical compound.

5.3.3 Inorganic Synthesis of Engineered Nanomaterials

Green Chemistry metrics are more rarely found to be applied to chemical synthesis or preparation procedures of inorganic chemical compounds, with respect to their application to the synthesis of organic compounds.

However, with the always more and more substantial advancements of nanotechnologies and related research areas, together with the lack of indisputable long-term environmental and human health effects of engineered nanomaterials, the possibility to benefit from quantitative and trustworthy Green Chemistry metrics, for the subsequent selection of the greener alternative for the preparation of a particular engineered nanomaterial, is highly desirable and will in all likelihood contribute to a more sustainable development of nanotechnology.

Particularly, the case study selected is referred to the chemical synthesis of anatase TiO_2 nanoparticles. The choice can be easily justified by the fact that titanium dioxide (mainly in the anatase polymorph) is the most studied and applied semiconductor and photocatalyst as a consequence of its unique physicochemical properties, and even more so when in the nanometer size range.

Particularly, the bottom-up hydrolytic sol–gel synthesis of anatase TiO_2 nanoparticles was assessed by applying both the LCA methodology and the software EATOS (Pini et al. 2015), to the reaction summarized in (5.6), at the basis of the procedure patented and employed by Colorobbia S.p.A. (Baldi et al. 2008), one of the most important Italian suppliers of chemicals for the building sector.

$$Ti(OiPr)_4 + 4H_2O \rightarrow TiO_2 + 2H_2O + 4iPrOH \quad (5.6)$$

Results of the LCA study allowed to identify the electric energy contribution necessary to mix the sol and the one necessary to mix the sol and triton X-100 as

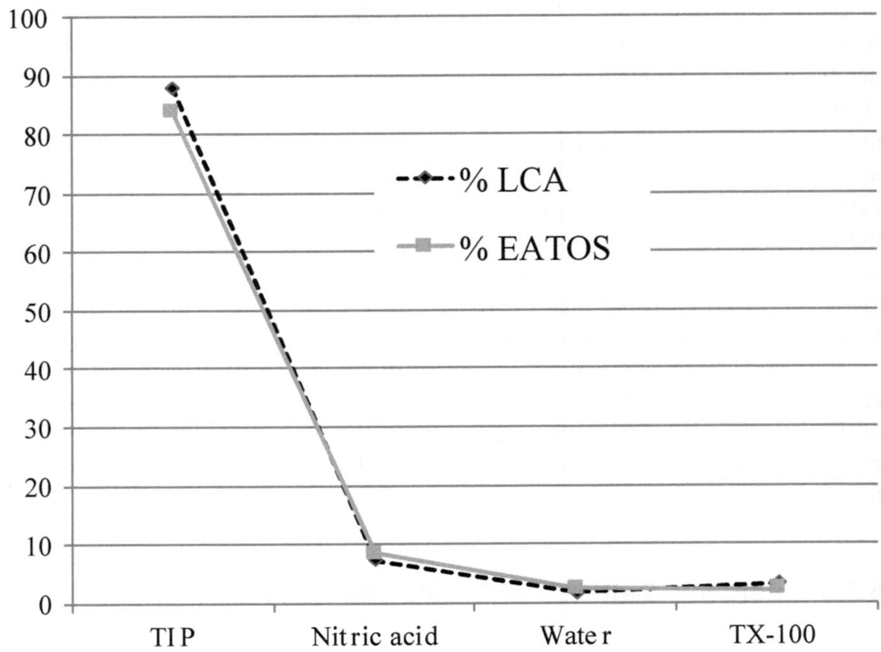

Fig. 5.8 EATOS and LCA (excluding energy consumptions) results comparison for the four most impacting contributions (TIP = Titanium tetraisopropoxide, TX 100 = Triton X 100). The percentages for the whole environmental impact are reported. Reproduced from Pini et al. (2015) with permission from The Royal Society of Chemistry

the most impacting factors affecting the overall preparation procedure. Therefore, according to the guidelines and recommendations given in the previous section, the application of EATOS software to perform similar evaluation should be discouraged. However, by completely neglecting any energy-related contribution during the life cycle impact assessment, the results obtained are indeed comparable with those obtained by simply applying the software EATOS, as reported in Fig. 5.8, thus contributing to strengthen the recommendation of the software EATOS as the less comprehensive metric/tool of choice to be applied in synergy with LCA methodology.

Conclusions

In this chapter, the complex issue of greenness of chemical synthesis or preparation procedures of chemical compounds performed at a laboratory scale has been addressed. Particularly, it has been highlighted that the use of LCA methodology also at a laboratory-scale research stage should be highly recommended, and in some cases, mandatory, although a wide variety of less comprehensive metrics and tools are freely available and of more simple application, with respect to LCA.

Some of the most widely applied metrics and tools have been shortly overviewed in the present chapter, in order to highlight their benefits, their disadvantages, and their application possibilities, particularly in synergy with the all-encompassing methodology represented by life cycle assessment.

Indeed, in order to partially reduce the typical laborious and highly time-demanding nature of an LCA study without of course affecting the trustworthiness of the results obtained at a laboratory scale, some strategies and guidelines have been proposed in this contribution, most of which comprises the synergic application of less comprehensive Green Chemistry metrics. Particularly, the software EATOS, and the metrics it allows calculating, has been identified as the tool of choice to be applied before or concurrently to the LCA evaluation, mainly as a consequence of the widely reported parallelism of its results with those obtained by LCA, which has been demonstrated in the preparation of both organic and inorganic compounds.

Exactly in this framework, some case studies have been presented and overviewed, comprising chemical synthesis of target organic compounds, their comparison with extraction of the desired molecule from natural resources, as well as the chemical synthesis of engineered oxide nanomaterials. In all of the presented case studies, both the software EATOS and a cradle to the grave life cycle analysis were applied, and the followed approach was justified on the basis of the previously proposed recommendations showing typically a high degree of synergy and coherence in the results among the two employed strategies.

References

ACS Green Chemistry Institute pharmaceutical roundtable 2018. https://www.acs.org/content/acs/en/greenchemistry/research-innovation/tools-for-green-chemistry.html#process-mass-intensity-calculator. Accessed 17 Apr 2019

Albini A, Protti S (2016) Green metrics, an abridged glossary. In: Paradigms in Green Chemistry and technology, Springer Briefs in Green Chemistry for Sustainability. Springer, p 11

Allen DT, Hwang BJ, Licence P et al (2015) Advancing the use of sustainability metrics. ACS Sustain Chem Eng 3:2359–2360

Anastas PT, Warner JC (1998) Green Chemistry: theory and practice. Oxford University Press, New York

Anastas PT, Zimmerman JB (2003) Design through the 12 principles of green engineering. Environ Sci Technol 37:94A–101A

Andraos J (2005) Unification of reaction metrics for Green Chemistry: application to reaction analysis. Org Process Res Dev 9:149–163

Andraos J (2009) Global Green Chemistry metrics analysis algorithm and spreadsheets: evaluation of the material efficiency performances of synthesis plans for oseltamivir phosphate (Tamiflu) as a test case. Org Process Res Dev 13:161–185

Andraos J (2016) Critical evaluation of published algorithms for determining material efficiency green metrics of chemical reactions and synthesis plans. ACS Sustain Chem Eng 4:1917–1933

Andraos J, Mastronardi ML, Hoch LB et al (2016) Critical evaluation of published algorithms for determining environmental and hazard impact green metrics of chemical reactions and synthesis plans. ACS Sustain Chem Eng 4:1934–1945

Armenta S, Garrigues S, de la Guardia M (2008) Green analytical chemistry. TrAC Trends Anal Chem 27:497–511

Baldi G, Bitossi M, Barzanti A (2008) US Patent 0317959 A1

Becker H (2001) Organikum: organisch-chemisches Grundpraktikum. Wiley-VCH, Weinheim

Bianchini R, Corsi M, Bonanni M (2014) WO Patent 2014177528A1

Caramazana P, Dunne P, Gimeno-Fabra M et al (2017) Assessing the life cycle environmental impacts of titania nanoparticle production by continuous flow solvo/hydrothermal syntheses. Green Chem 19:1536–1547

Caramazana P, Dunne P, Gimeno-Fabra M et al (2018) A review of the environmental impact of nanomaterial synthesis using continuous flow hydrothermal synthesis. Curr Opin Green Sustain Chem 12:57–62

Cespi D, Passarini F, Vassura I et al (2016) Butadiene from biomass, a life cycle perspective to address sustainability in the chemical industry. Green Chem 18:1625–1638

Chemat F, Vian MA, Cravotto G (2012) Green extraction of natural products: concept and principles. Int J Mol Sci 13:8615–8627

Dallinger D, Kappe CO (2017) Why flow means green—evaluating the merits of continuous processing in the context of sustainability. Curr Opin Green Sustain Chem 7:6–12

Do JL, Friscic T (2017) Mechanochemistry: a force of synthesis. ACS Cent Sci 3:13–19

Duarte RC, Ribeiro MGTC, Machado AASC (2015) Using green star metrics to optimize the greenness of literature protocols for syntheses. J Chem Educ 92:1024–1034

Ecoscale calculator website (2006) http://ecoscale.cheminfo.org/calculator. Accessed 17 Apr 2019

Eissen M, Metzger JO (2001) EATOS user manual website: http://www.metzger.chemie.uni-oldenburg.de/eatos/eatosmanual.pdf. Accessed 18 Apr 2001

Eissen M, Metzger JO (2002) Environmental performance metrics for daily use in synthetic chemistry. Chem Eur J 8:3581–3585

Eissen M, Metzger JO (2019) EATOS, environmental assessment tool for organic syntheses, Software website: http://www.metzger.chemie.uni-oldenburg.de/eatos/english.htm. Accessed 18 Apr 2019

Ferrari E, Pignedoli F, Imbriano C et al (2011) Newly synthesized curcumin derivatives: crosstalk between chemico–physical properties and biological activity. J Med Chem 54:8066–8077

Ferrari AM, Volpi L, Pini M et al (2019) Building a sustainability benchmarking framework of ceramic tiles based on Life Cycle Sustainability Assessment (LCSA). Resources 8:11

Freese U, Heinrich F, Rößner F (1999) Acylation of aromatic compounds on H-Beta zeolites. Catal Today 49:237–244

Hudlicky T, Frey DA, Koroniak L et al (1999) Toward a "reagent-free" synthesis. Green Chem 1:57–59

Hünig S, Märkl G, Sauer J (1979) Integriertes organisches praktikum. Verl Chemie, Weinheim

Leonelli C, Mason TJ (2010) Microwave and ultrasonic processing: now a realistic option for industry. Chem Eng Proc 49:885–900

Life+ Ecodefatting project website: http://www.life-ecodefatting.com. Accessed 19 Apr 2019

NOP (2019) Website: https://www.oc-praktikum.de/nop/en-article-why. Accessed 17 Apr 2019

Pabon HJJ (1964) A synthesis of curcumin and related compounds. Recl Trav Chim Pays-Bas 83:379–386

Pini M, Rosa R, Neri P et al (2015) Environmental assessment of a bottom-up hydrolytic synthesis of TiO_2 nanoparticles. Green Chem 17:518–531

Ravelli D, Dondi D, Fagnoni M et al (2010) Titanium dioxide photocatalysis: an assessment of the of the environmental compatibility for the case of the functionalization of heterocyclics. Appl Catal B 99:442–447

Ravelli D, Protti S, Neri P et al (2011) Photochemical technologies assessed: the case of rose oxide. Green Chem 13:1876–1884

Ribeiro MGTC, Machado AASC (2013) Greenness of chemical reactions—limitations of mass metrics. Green Chem Lett Rev 6:1–18

Ribeiro MGTC, Costa DA, Machado AASC (2010) "Green Star": a holistic Green Chemistry metric for evaluation of teaching laboratory experiments. Green Chem Lett Rev 3:149–159

Rosa R, Pini M, Neri P et al (2017) Environmental sustainability assessment of a new degreasing formulation for the tanning cycle within leather manufacturing. Green Chem 19:4571–4582

Rosa R, Ferrari E, Veronesi P (2018) From field to shelf: how microwave-assisted extraction techniques foster an integrated green approach. In: You KY (ed) Emerging microwave technologies in industrial, agricultural, medical and food processing. IntechOpen, London, pp 179–203

Sheldon RA (1992) Organic synthesis; past, present and future. Chem Ind (London): 903–904

Sheldon RA (1994) Consider the environmental quotient. ChemTech 24:38–47

Sheldon RA (2007) The E factor: fifteen years on. Green Chem 9:1273–1283

Sheldon RA (2018) Metrics of green chemistry and sustainability: past, present, and future. ACS Sustain Chem Eng 6:32–48

Stankiewicz A, Moulijn JA (2000) Process intensification: transforming chemical engineering. Chem Eng Prog 96:22–34

Vaccaro L, Lanari D, Marrocchi A et al (2014) Flow approaches towards sustainability. Green Chem 16:3680–3704

Van Aken K, Strekowski L, Patiny L (2006) EcoScale, a semi-quantitative tool to select an organic preparation based on economical and ecological parameters. Beilstein J Org Chem 2. https://doi.org/10.1186/1860-5397-2-3

Vogel AI (1978) Vogel's textbook of practical organic chemistry: including qualitative organic analysis. Longman, London

Zerazion E, Rosa R, Ferrari E et al (2016) Phytochemical compounds or their synthetic counterparts? A detailed comparison of the quantitative environmental assessment for the synthesis and extraction of curcumin. Green Chem 18:1807–1818

LCA as a Support Tool for the Evaluation of Industrial Scale-Up

6

Simone Maranghi, Maria Laura Parisi, Riccardo Basosi and Adalgisa Sinicropi

Abstract

The environmental impact associated with the various scale dimensions (i.e., laboratory, pilot plant, and industrial) is influenced by substantial changes correlated with many process parameters. Performing an industrial eco-design study starting from laboratory data is complex, and the risk of outlining an environmental profile that does not correspond to the real future industrial system is quite high. Usually, this is due to the scarcity of information about the industrial scale of the analysed system and to the difficulty of predicting the behaviour and evolution of the process during the scale-up. The purpose of this chapter is to highlight the main advantages and drawbacks of the application of LCA to support the industrial scale-up in the chemical sector. The matters addressed in the previous chapter will also be emphasised and integrated with other methodological issues. Moreover, an LCA methodological framework to deal with a systematic scale-up procedure overarching all the LCA phases is proposed.

S. Maranghi (✉) · M. L. Parisi · R. Basosi · A. Sinicropi
R²ES Lab (Research for Renewable Energy and Sustainability), Department of Biotechnology, Chemistry and Pharmacy, Italian MIUR Grant - Department of Excellence 2018-2022, University of Siena, Siena, Italy
e-mail: simone.maranghi@unisi.it

S. Maranghi · M. L. Parisi · R. Basosi · A. Sinicropi
CSGI (Center for Colloid and Surface Science), Florence, Italy

M. L. Parisi · R. Basosi · A. Sinicropi
Institute of Chemistry of Organometallic Compounds, (CNR-ICCOM), Florence, Italy

S. Maranghi and C. Brondi (eds.), *Life Cycle Assessment in the Chemical Product Chain*, https://doi.org/10.1007/978-3-030-34424-5_6

6.1 Introduction

The Life Cycle Assessment (LCA) of chemical products and processes can be performed at different scales of the production chain, from the very early laboratory scale through the pilot to the final industrial scale (Fantke and Ernstoff 2018). For each of these different development stages, a specific and suitable methodological LCA approach applies (Parisi et al. 2013). Thus, ambiguous results and misleading interpretations could be obtained if a comparison of environmental performances of systems at the different development stage is performed. A reliable LCA analysis can be performed if detailed, complete and reproducible data for the systems under study are available.

However, one of the most valuable applications of LCA is in the design phase of products and processes (namely eco-design). Indeed, the evaluation and assessment of future potential environmental impacts of a system in the early stages of development can allow for the identification of the critical hotspots before its effective large scale availability (Brezet and van Hemel 1997; Jeswiet and Hauschild 2005; Bravi et al. 2010). The results of an eco-design LCA can support the decision-makers and guide the technological and engineering development to get more sustainable products or processes (Galli et al. 2018).

An essential aspect of the design phase is the possibility to foresee the optimisation of systems at a large-scale already from the preliminary conception stage (Hetherington et al. 2014; Cucurachi et al. 2018). Indeed, most of the environmental impacts associated with the development of products or processes can be identified in the design phase (Tischner et al. 2000). Thus the LCA shows its particular useful in identifying and comparing the advantages and weaknesses of innovative processes or products with conventional ones (Shibasaki et al. 2006, 2007a). The application of LCA in the design phase involves the assessment of such processes on a small production scale that usually corresponds to laboratory processes. The system's environmental impacts calculated at such scale dimension can be used to project the assessment on a larger scale (Shibasaki et al. 2007b; Hetherington et al. 2014).

Performing an LCA of a chemical process developed at laboratory or pilot stage and scaling it up to predict its eco-profile at the industrial scale is a complex issue, that should be addressed step by step, by scaling-up each flow and process unit of the system under analysis. It is not uncommon that processes at the industrial scale are significantly different from their laboratory-scale counterpart and that the relationships among processes do not follow linear models (e.g., synergy effects) (Shibasaki et al. 2006). In addition to the different amount of material and energy flows involved, it is the entire chemical procedure that quite often changes radically, as it could be necessary to employ different equipment, reagents and chemical compounds other than those required in the laboratory. Predicting and modelling these changes in the development of an LCA are fundamental issues that can be accomplished thanks to a close collaboration between the LCA analyst and the industrial chemical sector expert.

In this context, many methodological and practical issues related to the description and definition of the parameters of the chemical process under study must be considered. Among these, the issue of a proper collection of data is essential to obtain a reliable dataset which may be used in the scale-up process. Data gathered in the Life Cycle Inventory (LCI) phase must be specific and representative of each unit process of the considered scale dimensions (Gavankar et al. 2015). Once these data have been comprehensively collected at one scale, they can be used to perform a reliable simulation of the process scale-up. It is important to note that the differences concerning the technologies employed, the raw materials used and their availability, the constraints imposed by laws and regulations affect the gathering of high-quality data directly, and thus the modelling of a reliable future industrial scenario could be problematic to perform (Piccinno et al. 2016). Moreover, the performance and efficiency of the chemical processes play a decisive role in affecting the environmental impact assessment obtained through the scale-up (Frischknecht et al. 2009).

Therefore, it is clear that, on the one hand, the application of LCA for eco-design is a compelling approach to foresee the environmental burdens in the very early development stage and for highlighting the hotspot and drawbacks related to the implementation of production processes at the industrial scale, in comparison with conventional and more mature products or processes. On the other hand, as shown in Fig. 6.1, this approach is characterised by a high degree of uncertainty, subjectivity and variability that should be minimised to obtain reliable results and substantial outcomes.

In this regard, recent updates concerning the prospective LCA approach could be very useful in order to harmonise the methodological framework for the evaluation of emerging systems and for setting up the scale-up procedure (Piccinno et al. 2016; Arvidsson and Molander 2017; Cucurachi et al. 2018; Hung et al. 2018). From a

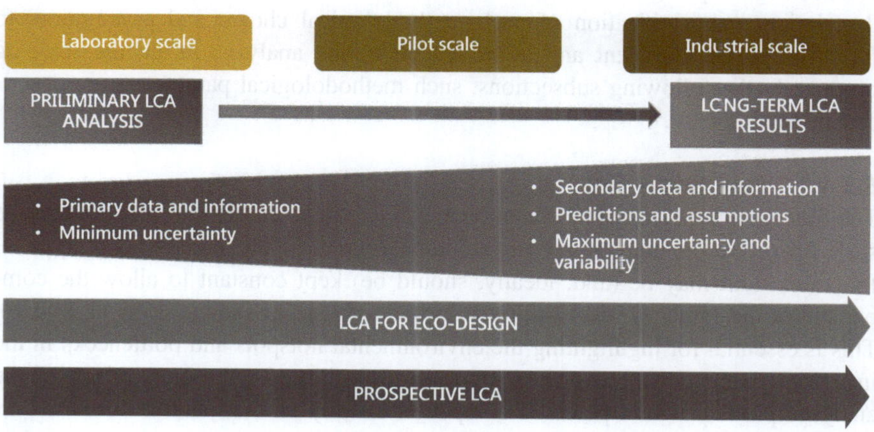

Fig. 6.1 Early-stage LCA: the synergies between eco-design and long-term prospective approach (Adapted from Hetherington et al. 2014)

methodological point of view, the application of prospective LCA to emerging technologies, production process or chemical procedure allows for the outlining of a scale-up framework for the evaluation of environmental burdens of future industrial scale, starting from data and information of laboratory or pilot scale.

In this chapter, several methodological approaches on the application of LCA to evaluate the potential environmental burdens of the industrial scale-up of chemical processes prospectively, are reviewed and discussed. The primary goal is to highlight the most relevant strength points and drawbacks of the proposed approaches while aiming at outlining a general framework for a reliable and thorough method to study the scale-up procedure from an LCA point of view.

6.2 Methodological Issues

Different methodological approaches concerning the application of LCA for the scale-up of chemical processes have already been published. (Shibasaki et al. 2006, 2007a, b; Caduff et al. 2011, 2012, 2014; Aboudheir et al. 2016; Piccinno et al. 2018). Before examining them (see Sect. 6.3), the most general methodological implications and critical issues that an LCA analyst should consider when performing the scale-up procedure of a chemical process are discussed below.

6.2.1 Scale-Up Procedure: Goal and Scope

The "goal and scope" phase is probably, together with the LCI, the LCA methodological step mostly affected by the change of the scale dimensions. The set-up of the methodological parameters of the analysis may differ substantially, considering the production scale. Particular attention should be given to the report, description and justification of each methodological choice and assumption, to facilitate the development and improvement of the analysis during the scale-up process. In the following subsections, such methodological parameters are described in detail.

6.2.1.1 Functional Unit

The functional unit should be comprehensively defined, taking into account that the scale-up procedure may influence the final output of the process. The type of functional unit may be (and, ideally, should be) kept constant to allow the comparison of the same process at different production scales (Shibasaki et al. 2007b). This is essential for highlighting the environmental hotspots and bottlenecks in the industrial production process and it would also allow for the comparison of the analysed product with a potential competitor already on the market.

When the functional unit is a specific function or a service and the chemical process under investigation is an innovative procedure performed in the laboratory, it might be problematic to predict which and how many services the chemical product will fulfil in a future industrial scenario (Hetherington et al. 2014). The LCA analysts should have to consider this issue by hypothesising several potential functional units and considering that the one selected could be reviewed in the interpretation and improvement phase of the analysis.

6.2.1.2 System Boundary

The definition of system boundary is strongly affected by the variations that occur moving from laboratory to industrial production scale. The number and typology of flows and unit processes could be completely different, and the definition of the appropriate system boundary and the comprehension of its modifications along the scale-up procedure is very complex. Shibasaki et al. (2006) try to provide a definition and classification of the most common development stages, starting from the very preliminary calculation models, up to the final industrial production scale (Fig. 6.2).

The industrial production stage is related to the features of a well-established process, that is ready for the market. All the other stages could be grouped and defined as "process development stages". These comprise the calculations models, the laboratory and mini plant, and the pilot plant. Among them, the laboratory and pilot scale are the most meaningful and widely investigated in the literature.

Laboratory procedures are the basis for understanding the behaviour of all the chemical reactions involved and for setting the most important parameters (i.e., amount of raw materials required, compounds molar ratio, reaction yield). On the other hand, reactions performed at pilot scale represent the last test and development step necessary for the finalisation and optimisation of the process before the final industrial production. The definition of an intermediate step could fulfil the gap between a process in a very-early-stage of development and the same process in the near industrial production. The differences between laboratory and/or pilot and industrial chemical production process are schematically shown in Fig. 6.3.

Indeed, while the laboratory procedure is primarily focused on obtaining the chemical product, an industrial production process consider several different parameters (e.g., reaction yield, process automatization, the human and environmental toxicity of raw materials and processes, the economic sustainability of the process) that inevitably lead to some changes in the system boundary definition.

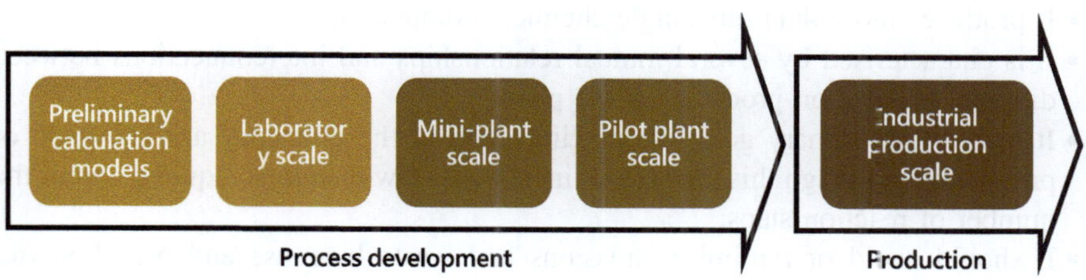

Fig. 6.2 Process development along with the different production scales (adapted from Shibasaki et al. 2006)

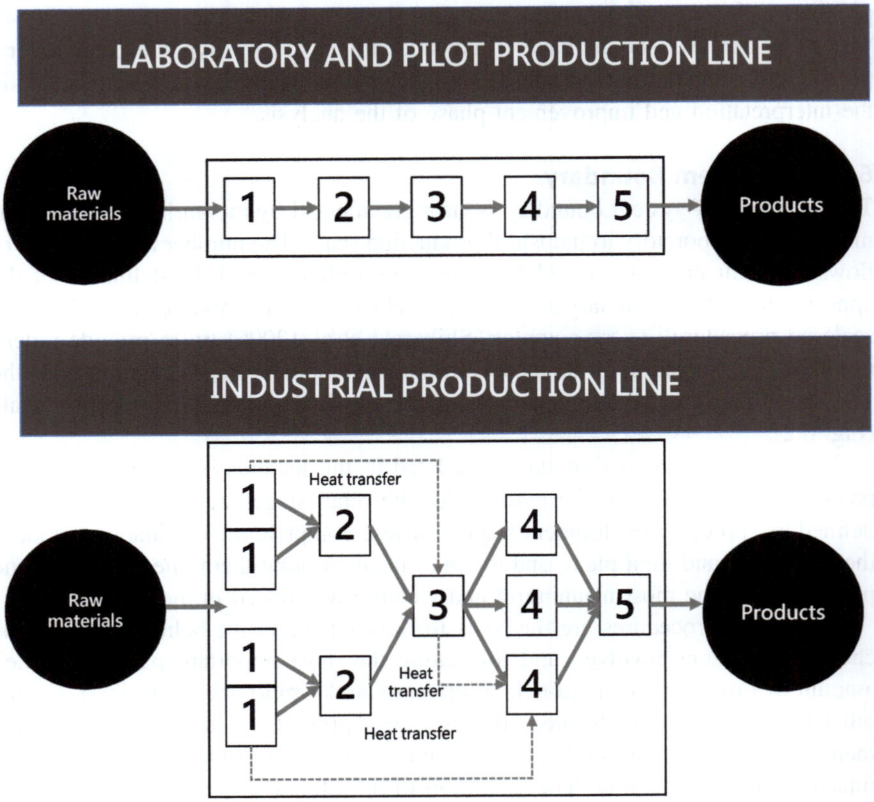

Fig. 6.3 Synergies and differences between laboratory/pilot and hypothetical industrial chemical production processes (Adapted from Shibasaki et al. 2007b)

Thus, considering such differences, it is necessary to consider a system expansion of the boundaries. Moreover, performing an LCA at laboratory or pilot scale, the LCA analyst should try to outline the system boundaries of the future industrial process, considering that, usually (Shibasaki et al. 2007a):

- It produces more than one single chemical compound;
- It is characterised by several mutual relationships and interconnections between different production processes at the plant;
- It has, as the primary goal, the maximisation of the efficiency and the yield of production, although this leads to an increase of raw materials required and of the number of reaction steps;
- It should avoid or minimise emissions and waste by reuse and recycling (respecting specific environmental regulation).

Besides the definition of the geographical and spatial system boundary, also the temporal range differs considerably when changing the production scale. Moving from the laboratory to pilot and to industrial dimension, the time window of the system should be expanded, developing specific future scenarios that take into account the possible lifetime of the product, its use and end-of-life phases, by implementing a long-term prospective LCA.

6.2.1.3 Allocation Procedure

System expansion could be implemented along with the scale-up procedure, allowing to avoid several problems concerning the choice of a suitable allocation procedure. However, the high probability that a future industrial chemical production process could have multiple output flows should lead the LCA analyst to decide which allocation procedure is most appropriate for the analysed system (Burgess and Brennan 2001). The most common allocation procedure is based on the mass, but the allocation method based on the number of moles, energy content, and economic value are used too. The use of economic allocation by the scientific researchers is increasing because it represents a valid criterion to distinguish products, co-products and waste, giving relevant information concerning the economic value of these output flows. It allows to allocate the impact based on the function of the product, rather than based on its physical characteristics (Ardente and Cellura 2012).

6.2.1.4 Data Requirement

This issue deals with data gathering and quality, which is a critical point in LCA (Parisi et al. 2019). The unavailability of primary data makes the use of secondary and literature data unavoidable (Maranghi et al. 2019). This problem is particularly relevant in the process scale-up of chemicals manufacturing, mainly due to the issues already mentioned above. Different methodological approaches, such as economy of scale, statistical simulations, learning curve and physicochemical relationship are proposed to overcome the problem of data lacking (Caduff et al. 2014; Piccinno et al. 2016; Cucurachi et al. 2018). More details concerning data gathering and the LCI phase are discussed in the next paragraphs.

6.2.1.5 Impact Assessment

The choice of impact categories or the selection of a suitable Life Cycle Impact Assessment (LCIA) method for the calculation of the environmental impact of chemicals is widely discussed in Chap. 1. For the scale-up procedure purpose, the choice of the proper LCIA method should be principally driven by the specific features of the chemical compound and its laboratory and industrial production chain. Moreover, the potential future environmental burden that could be related to the use and disposal phase should be considered too. The LCA analyst should select the impact assessment categories related to the whole projected life cycle of the industrial production scale. The identification of environmental hotspots is crucial for proper modelling of prospective scenarios, that should account for all the possible future impacts at the industrial production scale (Piccinno et al. 2018).

6.2.2 Life Cycle Inventory

In order to determine which procedures and technologies of the LCI of a process at the laboratory scale could be adapted at the industrial scale, expertise in the field of chemical engineering and industrial chemistry are required (Simon et al. 2016; Righi et al. 2018; Parvatker and Eckelman 2019).

The main challenge is represented by the identification of the potential evolution of each unit process and input and output flow that makes up the LCI, by supposing which industrial methods and techniques could be employed in the future production chain (Frischknecht et al. 2009).

Along with the development of a chemical process, there might be several changes in the LCI filling procedure. This is mainly due to:

- The employment of different raw materials (e.g., reactants, solvents, eluents);
- The use of a different source of energy (e.g., from electricity to thermal energy);
- The change of technologies and equipment employed;
- Opportunities for energy and material integration;
- Increased yield, efficiency, and productivity of the process;
- Multiple outputs (e.g., co-products and by-products);
- Qualitative and quantitative changes in emissions and waste generation;
- End-of-life phase (i.e., disposal, recovering, reuse and recycling processes);
- The design of the future chemical plant.

All these issues should be carefully analysed, following a systematic procedure leading to a specific scale-up framework for each of them Piccinno et al. (2018). Concerning the raw materials, the increase in their use is driven by stoichiometric reactions and the final yield of the chemical process. At the industrial scale, the minimisation of raw materials consumption coupled with the maximisation of the final product is essential along with the development of the process. To this aim, catalysts are widely used in order to provide a more efficient and performing pathway for the chemical process.

Concerning the yield of chemical procedures, the primary scope of a process at the laboratory production scale is to have a proof of concept of the desired final product. On the contrary, at the industrial scale, the main goal is the production of a large amount of the final product, while limiting as much as possible the costs in terms of materials and energy consumption. However, the final product should have specific commercial requirements, such as high purity or particular storage properties, for which additional production steps may be necessary. This could lead to an increase in raw materials and energy flows and a decrease in the final reaction yield (Piccinno et al. 2016). In addition to this, energy consumption is another crucial issue that needs to be faced during scale-up. Its reduction and the employment of thermal energy in place of electricity, represent crucial points along with industrial development.

Concerning the end-of-life phase, generally, for a chemical production process, a fair amount of hazardous waste is produced. At the industrial scale, this waste must be appropriately treated and, a large amount of it can be recycled and reused. Therefore, it is necessary to model a proper scenario analysis considering all the end-of-life routes and pathways of each chemical waste. For the considerable amount of solvent and eluents consumed during chemical processes, a recovery rate of about 68% is usually taken as a reference (Capello et al. 2005).

Concerning the modelling of emission scenarios, fugitive emission can occur in all the production stages, and proper measurement of all of them is quite problematic (see also Chap. 1). However, several methods allow for the estimation of fugitive emissions (Shine 1996; EIIP 2007; López-Aparicio et al. 2015), and it is essential to model these outputs because they can affect the environmental profile of the industrial production process crucially.

6.3 Practical Applications

In the following sections, the most relevant studies concerning the application of different methodological approaches for the scale-up of processes in the chemical sector are reported.

6.3.1 Physico-Chemical Screening

This approach comes from the results of a study on the systematisation of the scale factor in the LCA (Maranghi 2012). The main goal is the identification, within the building up of the LCI, of the *key elements* for the modelling of material and energy flows scaling-up and the understanding of the trend of the parameters that influence the environmental assessment in the scale-up. The physico-chemical screening focuses on the empirical observation of three different systems in order to extract the primary relationships that connect the processes in the change of scale dimensions.

The chosen processes for the analysis are (i) the organic synthesis of aniline, (ii) an innovative textile process, which combines foulard and plasma finishing techniques and (iii) a nano-silver functionalization process via plasma sputtering technique.

Data for the synthetic procedure of aniline at laboratory scale are taken from Furniss et al. (1989) while for the industrial synthesis are taken from the database Ecoinvent 3.01 (Ecoinvent Centre 2013). The synthesis of aniline turns out to be very dependent on the scale dimension. Furniss et al. (1989) report two different reactions, both having the nitrobenzene compound as the precursor. The following reduction of nitrobenzene to aniline depends strictly on the dimension of the scale. The high number of materials (reactants, catalysts and eluents) required in the laboratory synthesis is due to the step of purification by distillation, which is absent in the industrial process. In the large scale, the reduction process occurs in a single step using hydrogen. The reaction yield is very high for both cases (higher than 95%). The functional unit of the two processes is 1 kg of the final product.

The innovative finishing process is assessed with primary data available for the laboratory and the pilot scale (Fatarella et al. 2015; Parisi et al. 2013). Concerning the input and output material flows, the difference between the process at laboratory and pilot scale are minimal and they are concentrated only on the two finishing techniques: the foulard process and the plasma treatment (Busi et al. 2016).

The production process of self-cleaning t-shirts functionalized with nanomeric silver is assessed with secondary data from the literature (Walser et al. 2011) related to the laboratory, pilot and industrial scale. The plasma-sputtering process is characterised by a single step that is similar in all the considered scale (Walser et al. 2011). It operates directly on the finished tissue, so all the processes that are part of the conventional textile finishing are not included in the system boundaries. Moving from the laboratory scale through the pilot to the industrial production, a progressive decrease in the amounts of materials of the input and output flows is reported. This trend does not occur for the amount of silver (that varies slightly in the three scales) and for the nitrogen (which is used only in the pilot scale to fill the plasma sputtering reactor). Another difference is the presence of waste treatment in the industrial production stage. The chosen functional unit is the area of functionalized tissue.

In order to rationalise the scale-up process within the physico-chemical screening method, the analysis is divided into two stages: the first one concerns the modelling of direct energy consumption, and the second is the more complex modelling of the direct input of materials. The indicators used for the analysis are the Cumulative Energy Demand (CED), the Global Warming Potential in 100 years (GWP100) and the ReCiPe Endpoint. The calculations were performed with the software SimaPro 8.0.1, and the reference database is the Ecoinvent 3.01.

Comparing the small-scale production to larger ones, it is possible to observe that the contribution to the environmental profile of the raw materials usage is sensibly higher than the contribution of direct energy consumption. The resulted trend shows a difference of 10^{-1} magnitude in the transition from laboratory to pilot scale and a difference of 10^{-2} magnitude going from pilot to industrial scale (Fig. 6.4). This behaviour is independent of the type, the function and the steps that define the various processes in the considered scales.

In order to estimate the trend of the input and output of materials in the scale-up, some *key elements* for each case study are used as a reference. Figure 6.4 shows the results obtained with the indicators CED, GWP and ReCiPe calculated for the selected *key element* for each case study. These results are expressed using a parameter obtained from the following formula:

$$I = IND_{element}/IND_{total}(\%) \tag{6.1}$$

Equation 6.1 allows comparing the results obtained from different LCIA methods. The various *I* indexes are calculated as a percentage of the ratio between

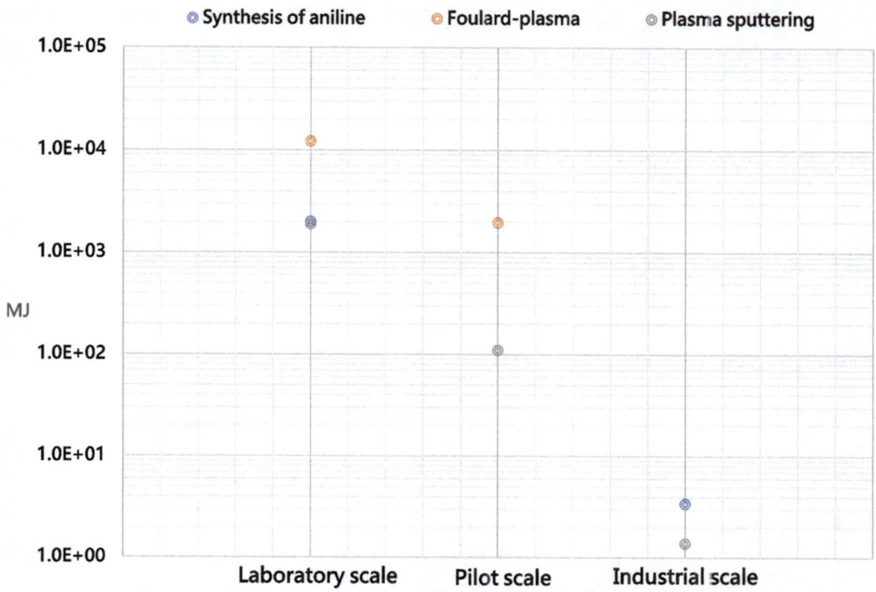

Fig. 6.4 Diagram in a logarithmic scale of direct energy consumption calculated with CED method

the calculated value of each input/output process ($IND_{element}$) and the total impact of the whole system with the same indicator (IND_{total}).

Analysing the *I* value for all input and output flows of the three systems, it is decided to examine in deeply these three *key elements*:

- nitrobenzene for the synthesis of aniline;
- methacrylate for the foulard-plasma process;
- silver for the plasma sputtering functionalization.

The environmental profile calculated with CED, GWP100, and ReCiPe of these three processes is shown in Fig. 6.5. The blue values refer to nitrobenzene, the green values refer to methacrylate, and the orange values refer to silver.

The rationalisation of the results obtained from the input and output of the materials during the scale-up is complex because of the considerable differences that can be found by comparing systems that operate on various scale dimensions. These differences mainly concern (i) the raw materials used, (ii) the choice of technologies used and (iii) the number of steps of the production process.

The case of the organic synthesis of aniline is the one that shows the most evident changes related to all three. In the case study of plasma-sputtering, there are variations related to two aspects: the raw materials used and the steps which set up the process (e.g., the reduction of emissions). The comparison of all the inputs and outputs of materials of the synthesis of aniline and the plasma sputtering process

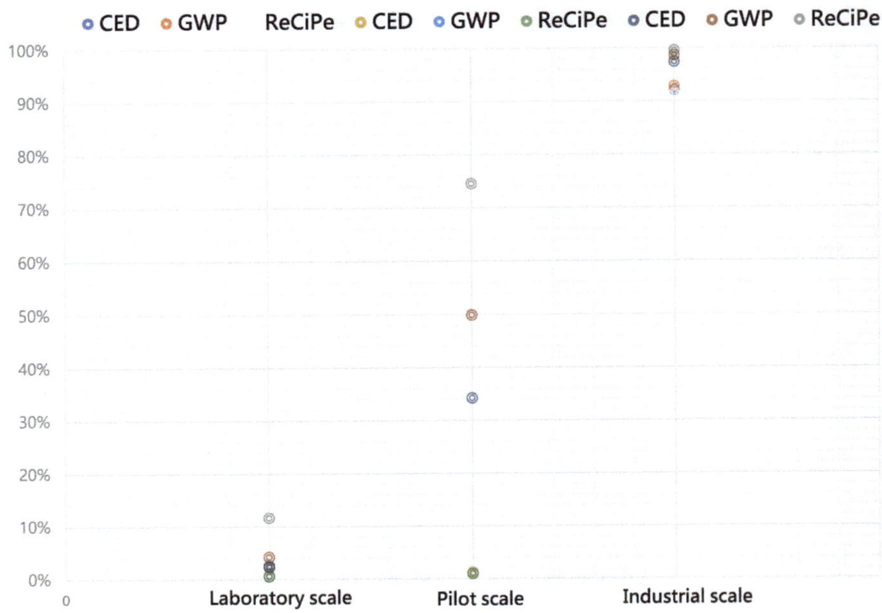

Fig. 6.5 The parameter *I* for the selected *key elements* calculated with CED, GWP100 and ReCiPe indicators, for synthesis of aniline (nitrobenzene, blue colour), textile finishing process (methacrylate, green colour) and plasma sputtering process (silver, orange colour)

highlights that the selected *key element* is the only one that has a reproducible trend in the scale-up. Therefore, it is possible to outline some specific characteristics that describe the *key element*:

- The relative amount per functional unit is constant in the transition from one scale to another;
- The environmental impact is high for all methods of calculation used;
- Is improbable and therefore it has high information content.

The concept of improbability derives from the second law of thermodynamics, whereby an unlikely event usually has high information content. Concerning the case study of the organic synthesis of aniline, the nitrobenzene is an uncommon input (improbable) because it is used as a specific precursor (high information content). On the other hand, others input of matter (e.g., hydrochloric acid, ether, hydrogen, and so on) are widely used in other organic syntheses and many other chemical processes (high probability, low information content). The same considerations may be applied to the silver used in the case study plasma-sputtering.

A difficulty in the identification of the *key element* may occur when it is hidden in the environmental profile of the industrial process because it is subjected to abatement or recycling operations. In this case, the real risk is to overestimate the environmental impacts related to the *key element* following the model described above.

6.3.2 Prognosis Method

Shibasaki et al. (2006, 2007a, b) proposed a method that uses the LCA results of a pilot scale process for identifying the most relevant aspects of the scale-up procedure. This so-called "prognosis method" allows to support the planning of industrial production, and to define the proper functional unit and system boundary along with the scale-up procedure. The interesting part of this approach is the application of the prognosis method to describe different "development levels" of the process and to investigate "synergy effects" that come out at the industrial production scale (Shibasaki et al. 2006). The starting point of this approach is the analysis of the LCA results calculated at the pilot scale, thanks to which it is possible to cut off the processes with a negligible environmental impact (Shibasaki et al. 2007b). This procedure should be supported by the knowledge of an expert in LCA and chemical processes in order to minimise ambiguous assumptions along with the scale-up.

Subsequently, the scale-up with the prognosis method takes into consideration all the changes in materials and energy consumptions, as well as the variation of several essential parameters (i.e. yield of reaction, process efficiency, industrial optimization, type of energy used, reuse of waste, etc.), related to all the upstream and downstream processes of the production phase (Shibasaki et al. 2007a, b).

The final step of the application of the prognosis method is focused on the industrial plant and, in particular, on the plant configuration (e.g., apparatus arrangement) that may significantly affect the LCA and the scale-up procedure. In this phase, all the parameters concerning the future industrial production and commercial productivity should be carefully considered. The expected manufacturing volume, the rearrangement of the production line, the synergies between the processes, the hypothetic changes that may occur due to the industrial boundaries, should be considered and described. Shibasaki et al. (2007b) report some optimising factors and a general formula to calculate the reduction of primary electricity consumption, considering the reduction of idling time between the processing of more than one product, not considering the time necessary to turn-on and shutting down the equipment.

6.3.3 Scaling Relationships

The method proposed by Caduff et al. (2011, 2012, 2014) is based on the estimation of essential properties of the system along with the scale-up procedure to support the definition of the system boundary, the choice of allocation procedure and the evaluation of data requirement. Thanks to a general non-linear power law ($y = a \cdot x^b$) and with a massive amount of LCI data, the authors were able to predict the behaviour of generic energy conversion equipment and to estimate several fundamental properties of the analysed system (Caduff et al. 2014). The application of this relationship allows for the scale-up of raw materials and energy consumption and costs of the most diffused equipment for energy conversion (Caduff et al. 2011). Thanks to the statistical method employed, it could be possible to quantify the magnitude of the changes in the LCI data (i.e., the amount of input and output flows) when the production scale increases considerably.

The scaling law, the estimated parameters, and key properties are specific for the investigated system. Thus, this approach should be fitted to the process. However, Caduff et al. demonstrate that the empirical scaling relationship law is suitable for the scale-up of a different system, e.g., common energy conversion equipment (Caduff et al. 2011), wind power plant (Caduff et al. 2012) and heat generation technologies (Caduff et al. 2014).

This method is suitable for the estimation of LCI flows, key properties, and impact assessment results. Thanks to the application of the scaling law to a broad set of specific data, peculiar scaling factors can be derived, and they can be used to model the scale-up, finding the appropriate industrial production condition of the process. Moreover, this method is an additional tool to support the design of prospective LCA, particularly for systems that maintain a similar outline and a similar function along with the technological development (Caduff et al. 2014).

6.3.4 Systematic Scale-Up Framework

The primary goal of the method proposed by Piccinno et al. (2016, 2018) is the identification and mathematical estimation of the environmental impacts of industrial chemical production, starting from primary data and information at the laboratory scale. This approach faces the problem of the significant difference among the production scales providing a systematic scale-up of every single process and flow that make up the LCI. Furthermore, this approach could be very useful to define the suitable system boundary, the allocation procedure, data requirement and impact assessment method.

Similarly to other approaches, the scale-up framework follows a step by step modelling that tries to outline and describe all the changes that occurs during the development of the system, starting from the laboratory chemical process, passing through the scale-up of input and output flows and the design of a simple industrial plant, up to the scale-up and linkage of the process steps (Piccinno et al. 2016). Useful information concerning how to manage common laboratory processes in order to predict which ones could be different in industrial production, and how the LCA analysts can model these processes, are reported. The authors describe in details the systematic framework for scaling-up several LCI flows, such as raw materials (e.g., reactants, solvents, catalysts), heating energy, chemical laboratory procedures (e.g., stirring, homogenizing, grinding, filtration and distillation, drying), waste and other output flows and infrastructure (Piccinno et al. 2016).

The application of this scale-up framework of LCI allows for the first estimation of the impact of the industrial chemical process. It can effectively support the evaluation of the practical industrial scalability of a chemical process at the laboratory research scale, identifying the main environmental hotspots that should be well-monitored during the future development of the process (Piccinno et al. 2018).

6.4 Advantages and Drawbacks of Methodological Approaches: Lesson Learned from Practical Application

All the efforts made by several researchers demonstrate that the scale-up modelling of chemical products and processes is within reach, although there is still further work to do to improve the proposed procedures from the methodological point of view.

Nevertheless, following the approaches and methods proposed in the literature and summarised in this chapter, it is already possible to outline a general methodological framework for the LCA scale-up of chemical processes, that is characterised by a different level of complexity and completeness (Fig. 6.6).

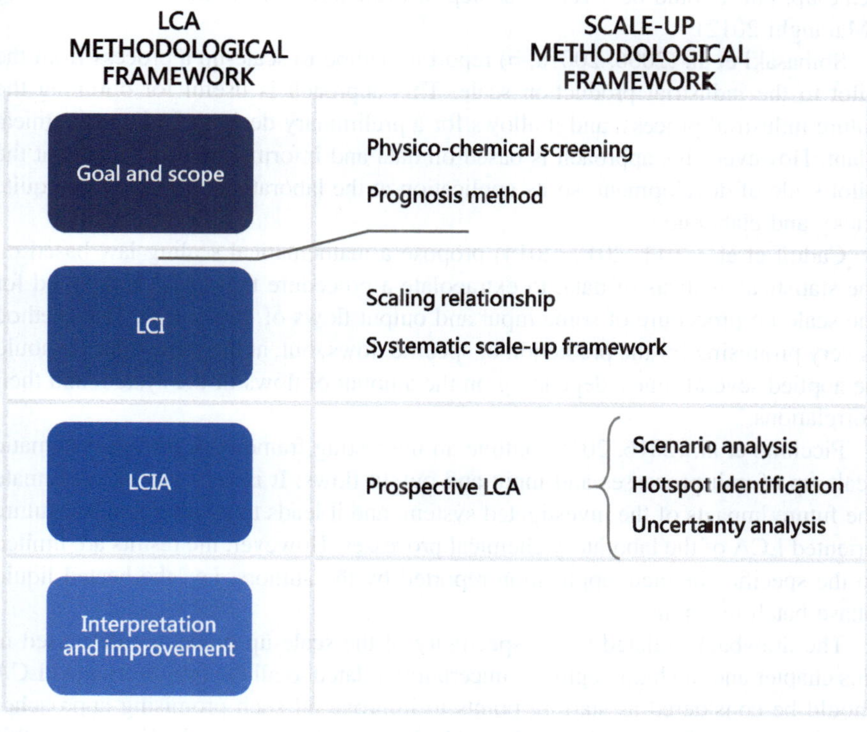

Fig. 6.6 Synergies and interdependencies between LCA and hypothetical overarching scale-up frameworks

All the LCA driven methodological approaches reported in the framework shown in Fig. 6.6 deal with the scale-up procedure with a slightly different point of view. The physico-chemical approach is mainly recommended for a preliminary screening of the process, the prognosis method is useful to start the modelling of the future industrial system, while the scaling law method and the systematic scale-up framework are mostly focused on the practical scale-up of the LCI data and further LCA calculation. Furthermore, the very recent effort put in the field by several researchers on the definition of the methodological approach for prospective LCA allows for a beneficial synergy for coupling the eco-design of the emerging system and the industrial scale-up (Arvidsson and Molander 2017; Arvidsson et al. 2018; Cucurachi et al. 2018). Thus, prospective LCA could be an essential methodological approach that allows for a reliable LCIA for eco-design of chemical processes.

All the methods and approaches reported in this chapter have advantages and drawbacks. Maranghi (2012) proposed a useful method for a very preliminary evaluation of the process, with the application of LCA at laboratory scale and the calculation of the *I* (i.e., the incidence of the *key element* on the overall environmental impacts), supported by the thermodynamic concepts. Thanks to this, it is possible to identify which could be the processes that need to be inspected for the future industrial environmental assessment. Thus it is not a practical model for scale-up, but it could be a very first step of evaluation for the following modelling (Maranghi 2012).

Shibasaki et al. (2006, 2007a, b) report a method to scale-up a process from the pilot to the industrial production scale. This approach is useful for outlining the future industrial process, and it allows for a preliminary design of a large chemical plant. However, this approach is based on data and information of a process at the pilot scale of development, so its application at the laboratory stage could be quite tricky and elaborate.

Caduff et al. (2011, 2012, 2014) propose a mathematical scaling law based on the statistical analysis of data, to extrapolate a procedure that could be applied for the scale-up procedure of some input and output flows of the system. This method is very promising for the prediction of specific flows, but, at the same time, it should be applied several times, depending on the amount of flows of the system and their correlations.

Piccinno et al. (2016, 2018) outline an interesting framework for the systematic scale-up of unit processes and input and output flows. It is very useful to estimate the future impacts of the investigated system, and it leads to a rather realistic future oriented LCA of the laboratory chemical processes. However, the results are limited to the specific chemical application reported by the authors, i.e., the heated liquid phase batch reactions.

The drawbacks related to the specificity of the scale-up procedure discussed in this chapter and the high degree of uncertainty related to all the future-oriented LCA should be considered as starting points to improve all such promising approaches and the general framework for the scale-up procedure shown in Fig. 6.6 In this context, the application of prospective LCA, which tries to consider all the changes that occur during the industrial development of a system, should be a suitable framework for improving the scale-up procedure. The prospective approach already reported in the literature (Frischknecht et al. 2009; Hetherington et al. 2014; Arvidsson et al. 2018; Cucurachi et al. 2018), allows for the assessment of the LCIA with the strengthening of scenario analysis, hotspots identification, and uncertainty calculation. Nevertheless, the prospective LCA (and the method reported in this chapter to address it) should also be focused on the modelling of goal and scope, LCI and LCIA phases of LCA.

Conclusions

This chapter aims to report and summarise the most important methodological issues concerning the application of LCA as a support tool for the industrial scale-up procedure of chemical processes. In particular, it deals with the identification of the advantages and drawbacks of each method, in order to make an overall evaluation of them and to suggest a preliminary framework for a systematic scale-up procedure that involves all the LCA phases. The main methodological issues that should be considered when implementing LCA for the eco-design of chemical processes to support their industrial development are identified.

All the reported methods and frameworks are characterised by a high degree of uncertainty due to the numerous different factors that affect the development of chemical processes. Indeed, the application of scale-up in the chemical sector is a complex, long and case-specific procedure, and it requires high qualified knowledge of chemistry, chemical engineering and robust experience in the application of LCA applied to the chemical production chain.

All these approaches need to be further developed and improved in order to extend their applicability to all the chemical industrial sector. However, the methodological approaches reported in this chapter have already demonstrated their soundness and constitute the basis for the development of LCA as a reliable tool that can support the industrial scale-up in the chemical sector.

References

Aboudheir A, Akande A, Idem R et al (2016) Life cycle assessment (LCA) applied to the process industry: a review. J Clean Prod 17:1028–1041. https://doi.org/10.1007/s11367-012-0432-9

Ardente F, Cellura M (2012) Economic allocation in life cycle assessment: the state of the art and discussion of examples. J Ind Ecol 16:387–398. https://doi.org/10.1111/j.1530-9290.2011.00434.x

Arvidsson R, Molander S (2017) Prospective life cycle assessment of epitaxial graphene production at different manufacturing scales and maturity. J Ind Ecol 21:1153–1164. https://doi.org/10.1111/jiec.12526

Arvidsson R, Tillman AM, Sandén BA et al (2018) Environmental assessment of emerging technologies: recommendations for prospective LCA. J Ind Ecol 22:1286–1294. https://doi.org/10.1111/jiec.12690

Bravi M, Parisi ML, Tiezzi E, Basosi R (2010) Life cycle assessment of advanced technologies for photovoltaic panels production. Int J Heat Technol 28:133–139

Brezet H, van Hemel C (1997) Ecodesign: a promising approach to sustainable production & consumption. United Nations Environment Programme, Industry and Environment, Cleaner Production

Burgess AA, Brennan DJ (2001) Application of life cycle assessment to chemical processes. Chem Eng Sci 56:2589–2604. https://doi.org/10.1016/S0009-2509(00)00511-X

Busi E, Maranghi S, Corsi L, Basosi R (2016) Environmental sustainability evaluation of innovative self-cleaning textiles. J Clean Prod 133. https://doi.org/10.1016/j.jclepro.2016.05.072

Caduff M, Huijbregts MAJ, Althaus H, Hendriks JA (2011) Power-law relationships for estimating mass, fuel consumption and costs of energy conversion equipments. 751–754. https://doi.org/10.1021/es103095k

Caduff M, Huijbregts MAJ, Althaus HJ et al (2012) Wind power electricity: the bigger the turbine, the greener the electricity? Environ Sci Technol 46:4725–4733. https://doi.org/10.1021/es204108n

Caduff M, Huijbregts MAJ, Koehler A et al (2014) Scaling relationships in life cycle assessment: the case of heat production from biomass and heat pumps. J Ind Ecol 18:393–406. https://doi.org/10.1111/jiec.12122

Capello C, Hellweg S, Badertscher B, Hungerbühler K (2005) Life-cycle inventory of waste solvent distillation: statistical analysis of empirical data. Environ Sci Technol 39:5885–5892. https://doi.org/10.1021/es048114o

Cucurachi S, Van Der Giesen C, Guinée J (2018) Ex-ante LCA of emerging technologies. Proc CIRP 69:463–468. https://doi.org/10.1016/j.procir.2017.11.005

EIIP (2007) Methods for estimating air emissions from chemical manufacturing facilities. In: EIIP EIIP- (ed). US EPA

Fantke P, Ernstoff A (2018) LCA of chemicals and chemical products. Life cycle assessment. Springer International Publishing, Cham, pp 783–815

Fatarella E, Parisi ML, Varheenmaa M, Talvenmaa P (2015) Life cycle assessment of high-protective clothing for complex emergency operations. J Text Inst 106:1226–1238. https://doi.org/10.1080/00405000.2014.985881

Frischknecht R, Büsser S, Krewitt W (2009) Environmental assessment of future technologies: how to trim LCA to fit this goal? Int J Life Cycle Assess 14:584–588. https://doi.org/10.1007/s11367-009-0120-6

Furniss BS, Hannafort AJ, Smith PWG, Tatchell AR (1989) Textbook of practical organic chemistry, fifth edn. VOGEL's, New York

Galli F, Pirola C, Previtali D et al (2018) Eco design LCA of an innovative lab scale plant for the production of oxygen-enriched air. Comparison between economic and environmental assessment. J Clean Prod 171:147–152. https://doi.org/10.1016/j.jclepro.2017.09.268

Gavankar S, Suh S, Keller AA (2015) The role of scale and technology maturity in life cycle assessment of emerging technologies: a case study on carbon nanotubes. J Ind Ecol 19:51–60. https://doi.org/10.1111/jiec.12175

Hetherington AC, Borrion AL, Griffiths OG, McManus MC (2014) Use of LCA as a development tool within early research: challenges and issues across different sectors. Int J Life Cycle Assess 19:130–143. https://doi.org/10.1007/s11367-013-0627-8

Hung CR, Ellingsen LAW, Majeau-Bettez G (2018) LiSET: a framework for early-stage life cycle screening of emerging technologies. J Ind Ecol 00:1–12. https://doi.org/10.1111/jiec.12807

Jeswiet J, Hauschild M (2005) EcoDesign and future environmental impacts. Mater Des 26:629–634. https://doi.org/10.1016/j.matdes.2004.08.016

López-Aparicio S, Vogt M, Vo DT (2015) Methods for estimating air pollutant emissions—Norsk institutt for luftforskning (NILU)

Maranghi S (2012) Energy efficiency optimization study of micro-generation and energy saving systems by LCA analysis. Dissertation, University of Siena

Maranghi S, Parisi ML, Basosi R, Sinicropi A (2019) Environmental profile of the manufacturing process of perovskite photovoltaics: harmonization of life cycle assessment studies. Energies 12:3746

Parisi M, Maranghi S, Sinicropi A, Basosi R (2013) Development of dye sensitized solar cells: a life cycle perspective for the environmental and market potential assessment of a renewable energy technology. Int J Heat Technol 31:143–148. https://doi.org/10.18280/ijht.310219

Parisi ML, Ferrara N, Torsello L, Basosi R (2019) Life cycle assessment of atmospheric emission profiles of the Italian geothermal power plants. J Clean Prod 234:881–894. https://doi.org/10.1016/j.jclepro.2019.06.222

Parvatker AG, Eckelman MJ (2019) Comparative evaluation of chemical life cycle inventory generation methods and implications for life cycle assessment results. ACS Sustain Chem Eng 7:350–367. https://doi.org/10.1021/acssuschemeng.8b03656

Piccinno F, Hischier R, Seeger S, Som C (2016) From laboratory to industrial scale: a scale-up framework for chemical processes in life cycle assessment studies. J Clean Prod 135:1085–1097. https://doi.org/10.1016/j.jclepro.2016.06.164

Piccinno F, Hischier R, Seeger S, Som C (2018) Predicting the environmental impact of a future nanocellulose production at industrial scale: application of the life cycle assessment scale-up framework. J Clean Prod 174:283–295. https://doi.org/10.1016/j.jclepro.2017.10.226

Righi S, Baioli F, Dal Pozzo A, Tugnoli A (2018) Integrating life cycle inventory and process design techniques for the early estimate of energy and material consumption data. Energies 11. https://doi.org/10.3390/en11040970

Shibasaki M, Warburg N, Eyerer P (2006) Upscaling effect and life cycle assessment. Lce 2006:61–64

Shibasaki M, Albrecht S, Kupfer T (2007a) Small scale and large scale plants—effect on life cycle assessment. In: 17th European symposium on computer aided process engineering—ESCAPE17 1–6. https://doi.org/10.1007/978-1-4614-6570-6_6

Shibasaki M, Fischer M, Barthel L (2007b) Effects on life cycle assessment—scale up of processes. In: Advances in life cycle engineering for sustainable manufacturing businesses, pp 377–381. https://doi.org/10.1007/978-1-84628-935-4_65

Shine B (1996) Methods for estimating volatile organic compound emissions from batch processing facilities. J Clean Prod 4:1–7

Simon B, Bachtin K, Kiliç A et al (2016) Proposal of a framework for scale-up life cycle inventory: a case of nanofibers for lithium iron phosphate cathode applications. Integr Environ Assess Manag 12:465–477. https://doi.org/10.1002/ieam.1788

Tischner U, Rubik F, Schmincke E, Prosler M (2000) How to do ecodesign? A guide for environmentally and economically sound design, Art Books

Walser T, Demou E, Lang DJ, Hellweg S (2011) Prospective environmental life cycle assessment of nanosilver T-shirts. Environ Sci Technol 45:4570–4578. https://doi.org/10.1021/es2001248

7 Phosphate Recovery from Exhausted Extinguishing Powders: A Case Study of Circular Economy in the Chemical Industry

Giovanni Dotelli and Emanuela Viganò

Abstract

Nowadays in Europe, an exhausted extinguishing powders (EEP) industrial recovering process is still missing, fertilizer demand is increasing, phosphorous is a critical raw material. In this perspective, an EEP treatment pilot plant was realized, allowing the recovery of a high-value, non-renewable raw material, phosphate, transformed into fertilizers. This case study offers an example of circular economy and industrial symbiosis in the chemical industry (EEP are valorised as secondary raw materials in another sector, otherwise disposed of as special waste and not recovered) and highlights how to tackle with chemical processes using waste as secondary raw material. The boundaries choice between first and second life and on how to assess processes comparative analyses are the main critical points to deal with. The innovative PHOSave process is based on a mechanical treatment, a washing phase with an aprotic solvent and a biological treatment. To evaluate its environmental feasibility an LCA study was performed at the design stage. The scenario considered is from cradle-to-gate, from the collected EEP to the micro-fertilizer produced. Life cycle impact assessment of the innovative PHOSave process, using the CML impact method, has demonstrated that solvent choice for the washing phase has a huge influence on the overall environmental performance; in any case, the highest burden comes from the granulation phase.

G. Dotelli (✉)
Materials and Chemical Engineering Department "Giulio Natta", Politecnico di Milano - Chemistry, Milan, Italy
e-mail: giovanni.dotelli@polimi.it

E. Viganò
CESI S.p.A., Milan, Italy

S. Maranghi and C. Brondi (eds.), *Life Cycle Assessment in the Chemical Product Chain*, https://doi.org/10.1007/978-3-030-34424-5_7

7.1 Introduction

The extinguishing powders (EP) are probably the most common extinguishing agents and are classified on fire classes in the following categories of "ABC" Polyvalent Powder, and "BC" Powder or Special Powders. The "ABC" polyvalent powders are the most widespread in the market and are composed of mono-ammonium phosphate (MAP), ammonium sulfate (in various concentrations), colouring additives, fluxing agents, and a silicone-based additive to make the powder water-repellent. The main sources of raw materials (such as mono-ammonium phosphate) is from outside Europe (in particular from Russia and North Africa). These raw materials are also used in agriculture (such as fertilizers) due to their specific nature of releasing nitrogen and phosphorus. The transformation process of the raw material in ABC extinguishing powders involves the micro-grinding (pulverization) of phosphate and sulfate, and the addition of additives (such as silicone oils) to ensure flowability and water-repellent features. This also makes the recovery of "waste dry powder" very critical in that it minimizes the possibility of ennobling raw materials such as phosphate.

One of the main goals of the PHOSave process, developed by an Italian SME, is to achieve a solution to the problem of exhausted extinguishing powders (EEP) allowing the recovery of a high-value, non-renewable raw material (phosphate) in an almost pure form. This aim is accomplished through an innovation process that removes silicone from EEP via a treatment involving solvents and magnetic separation.

Exhausted EP (EEP) are classified as special wastes due to the presence of heavy metals and additives, and the very fine particles that compose the powders (99% are under 0.250 mm, with at least 40% under 0.040 mm). In Europe, a real industrial recovering process of extinguishing powders is not yet in place, and this fact raises issues concerning pollution control and environmental safety. Indeed, currently some farms distribute the extinguishing powders directly onto the field via a manure spreader.

Illegal processes of disposal or reuse of exhausted EPs are currently happening in all of Europe. In fact, often the maintenance company staff illegally disposes of and/or reuse exhausted extinguishing powders resulting in enormous environmental damages. Further, legal obligations (i.e. the replacement of the extinguishing agents every 36 months) are often disregarded with serious consequences for social safety and producers as well; indeed, they sell less material than what would be necessary for correct human safety and market efficiency. This adversely affects maintenance companies also. Trade association official data show that in Italy there are about 18 million kilograms of dry exhausted powder not regularly disposed of. This is in part because of the prohibitive costs (a good part of which is due to transport). The estimates at the EU level indicate that 36 million kilograms should be the annual quantity of exhausted extinguishing powders to be treated and disposed of ESPP (2017). However, the new EU Fertilizing Productions Regulation (FPR), which is now finally published in the EU Official Journal of 25th June (EU 2019), opens the

European market for recycled fertilizers and hopefully will put an end to illegal EEP disposal or reuse processes.

In the present chapter, the market relevance of the fertilizer production industry is sketched to highlight the urgent need of saving primary phosphorous whose reserves are limited. To accomplish this goal is vital to devise economic and environmentally sustainable recovery processes of phosphorous from waste streams.

Seeing the annual volume production of EEPs, their recovering could be a valid substitute for fertilizers from primary raw materials. An extensive literature on LCA of fertilizers is here reviewed for comparison purposes. Then, the LCA evaluation of the new process is discussed in some detail, and finally, a critical perspective commented.

7.2 Fertilizers' Production Industry

To meet the growing demand for crops, more fertilizer will be required and in an improved nutrient balance. After a drop in 2015/16 (by about 1.0%) to 181 million tons (Mt) nutrients, world fertilizer demand is anticipated to recover in 2016/17 (+2.9%) to 186 Mt (Heffer and Prud'homme 2016).

The contraction in fertilizer demand of 2015/16 has been caused by the economic slowdown in many emerging and developing countries, by persistently low international prices for most agricultural commodities, and by dry conditions across South Asia, Southeast Asia, Latin America and Africa. According to the baseline scenario, the IFA (International Fertilizer Industry Association) stated that world demand would rise on average by 1.6% per year between the base year (average of 2013/14 to 2015/16) and 2020/21 and assuming average weather patterns, global fertilizer demand forecast to remain slightly below 200 Mt by 2020/21 (Heffer and Prud'homme 2016).

Potash is projected to have the highest growth rate (approximately 3% annual growth); phosphate consumption is expected to grow by 2.0–2.5% annually, while nitrogen consumption is expected to grow at less than 2.0% per year (PotashCorp 2014). The most recent estimates (www.wroldfertilizer.com) state that the global fertilizer market is expected to value at US$155.8 billion in 2019, and it is expected to register a Compound Annual Growth Rate CAGR of 3.8% during the forecast period (2019–2024).

Developing countries in Asia and Latin America account for almost two-thirds of global consumption. Potash consumption is distributed amongst a number of major regions, while nitrogen and phosphate use is more heavily concentrated in China and India. Grains and oilseeds account for nearly two-thirds of global nitrogen and phosphate consumption. Global potash consumption strongly depends on the crop to be fertilized. Oilseeds, fruits and vegetables account for almost 40% of potash use, similar to the amount used by grains.

The world's biggest suppliers of fertilizers are distributed all over the world: there is no country monopolizing the fertilizer industry. PotashCorp is the world's largest fertilizer producer and is one of the three global producers that manufacture all three primary crop nutrients. Mosaic is one of the leaders in phosphate and potash fertilizer production. Then, there are Uralkali and Belaruskali that focus on potash fertilizer production and exportation.

Concerning the phosphate outlook, Heffer and Prud'homme (2016) pointed out that global phosphate rock supply would grow by 11% compared with 2015 demand, to reach 250 Mt of phosphate concentrate in 2020. Africa, Saudi Arabia and China would account altogether for three-fifths of this 25 Mt increase. Global phosphoric acid capacity in 2020 is projected to expand by 13% over 2015, starting from 65.3 Mt P_2O_5 in 2020. Global capacity for the main processed phosphate fertilizers would grow by 7 Mt P_2O_5 between 2015 and 2020, up to 52 Mt P_2O_5. Three exporting countries (Morocco, Saudi Arabia and China) would account for the bulk of the increase (Heffer and Prud'homme 2016).

7.3 Literature Review

7.3.1 LCA of Extinguishing Powders

Unfortunately for Companies that want to prepare an Environmental Product Declaration (EPD) there is no Product Category Rule (PCR) both regarding the treatment of exhausted extinguishing powders and extinguishers. Furthermore, there is a lack of LCA studies on EEP and extinguishers; here are presented some case studies dealing with these themes, even if they are not properly focused on extinguishing powders. In 1997 Johnson et al. (1997) presented a case study of chlorofluorocarbons (CFC) replacement for fire extinguishers with a TEWI (Total Equivalent Warming Impact) analysis for a production sequence, that leads from chloroform to tetrafluoroethylene and then to HFC-227ea, a fire extinguishing agent. TEWI is defined as a life cycle approach for comparing global warming impacts; it was invented by the chlorofluorocarbons (CFC) industry to compare replacements for its products (in refrigeration and solvent applications), whose production has been banned under the Montreal Protocols. However, many TEWI studies omit what can be a significant part of the system, i.e. production. The study by Johnson et al. (1997) demonstrated the significance of production in the TEWI of a commercial industry sector of gaseous fire extinguishers. Indeed, almost one-third of the TEWI occurs in the production of the alternative and it is even greater than the TEWI contribution connected with its use and eventual disposal.

In the field of fire extinguishers and fire in general, there are also efforts on assessing life cycle aspects of the fire performance of a product that is not usually accounted for in the standard LCA methodology, such as fires, and the risks associated with them. Probably, in future studies of these products fire performance could be an important parameter to be taken under consideration. For example,

Simonson et al. (2005) tested a Fire-LCA methodology in three case studies: TV, cables and furniture.

It is important to point out also the presence of studies that analyse the possible use of EEP in other industrial activities; for instance, Praticò et al. (2010) tested the potential of fire extinguisher powder as a filler in bituminous mixes, and they said also that the powder could be used for fertilizer synthesis.

7.3.2 LCA of Fertilizer Production and Use

7.3.2.1 State of the Art

In the literature, there is a far larger number of studies dealing with the life cycle assessment of fertilizers. In 2014, Skowrońska and Filipek prepared a review paper on LCA studies related to the production and use of mineral fertilizers (Skowrońska and Filipek 2014). These studies are reported in Table 7.1, where the methodological choices of each one are specified. More recent LCA studies and those not reviewed by Skowrońska and Filipek (2014) are reported in Table 7.2.

As can be seen from Tables 7.1 and 7.2, these studies differ substantially in methodological choices, a fact that is inevitably entailing a difficult comparison among results of the life cycle impact assessment phase.

Most of the studies are defined as "cradle-to-field" as they consider the impacts coming from the production and also the spreading on the field. When the application of fertilizers in agroecosystems is not accounted for, the analysis is termed "cradle-to-gate", i.e. from the raw materials up to the industrial fertilizer production. When LCA is used in agriculture, the functional unit most often chosen is the weight of the raw material or product (e.g. 1 kg or 1 ton) or surface area (e.g. 1 ha). However, some authors recommend using these units simultaneously. According to Charles et al. (2006), in assessing the efficiency of a production system for a particular crop, the functional unit should be a ton of grain, whereas the hectare should be used in analysing production intensity. The possibility of using waste as secondary raw material for fertilizer production is analysed only by Linderholm et al. (2012) and Chiew et al. (2015). Some authors emphasize that LCA of fertilizers in crop production should take into account the larger scale of the system, including factors such as the quality of the yield, biodiversity and the multifunctionality of agroecosystems (Skowrońska and Filipek 2014); on the other hand this fact would imply huge efforts during the LCA study implementation.

7.3.2.2 Available Data in Commercial Databases

In most databases used to perform LCA studies, fertilizers have an important role. In particular, one of the most accredited databases for chemicals is Ecoinvent (https://www.ecoinvent.org/) and in this database a good number of data on production of organic and inorganic fertilizers are available. Concerning the inorganic fertilizers, and specifically phosphate fertilizers, data on production of mono-ammonium phosphate (MAP) and di-ammonium phosphate (DAP) are available for Europe (RER) and the rest of the World (Global production minus

Table 7.1 LCA studies reviewed by Skowrońska and Filipek (2014)

References	Area of the study	Functional unit (FU)	Impact categories[a]	System boundaries
Brentrup et al. (2001)	Different N fertilizers (CAN, UAN, urea) for a sugar beet production system (Germany)	1 t of extractable sugar	GWP, EP, AP, summer smog; Eco-indicator 95 (normalization and weighting)	Cradle-to-field
Brentrup et al. (2004)	Different N application rates for winter wheat production (UK)	1 t of grain	Land use, depletion of abiotic resources, GWP, HTP, AP, EP	Cradle-to-field
Charles et al. (2006)	Optimization of N, P and K fertilization intensity (Switzerland)	1 ha 1 t of grain produced 1 t of grain with 13% protein	Land use, energy consumption, GWP500, POCP, EP, AP, aquatic ECOTOX, terrestrial ECOTOX, HTP	Cradle-to-field
Ahlgren et al. (2010)	Using biogas instead of natural gas in the production of nitrogen fertilizers (Sweden)	1 kg of nitrogen, as ammonium nitrate (3 kg ammonium nitrate with 33.5% N), at the exit gate of the production facility	Land use, GWP, NREU, EP, AP; IPCC (2001)	Cradle-to-gate
Nemecek et al. (2011)	Environmental impacts of extensive farming (Switzerland)	Hectare and year, kg dry matter, Swiss Franc return	NREU, GWP, summer smog, EP, AP, terrestrial ECOTOX, aquatic ECOTOX, biodiversity, soil quality; EDIP97, CML01, IPCC 2001	Cradle-to-field
Linderholm et al. (2012)	Different phosphorus fertilizers: mineral fertilizer, sewage sludge, struvite precipitated from wastewater, phosphorus recovered from sludge incineration (Sweden)	11 kg P/ha (25.2 kg P_2O_5): the average phosphorus output-removal with harvest per hectare from Swedish farmland in 2007	GWP, EP, energy demand, cadmium flows to farmland	Cradle-to-field
Tuomisto et al. (2012)	Organic, conventional and integrated farming systems (UK)	1 t of winter wheat with 86% dry matter content	GWP, energy use, land use	Cradle-to-field

[a]DAP = diammonium phosphate; MAP = mono-ammonium phosphate; FMP = fused magnesium phosphate; TSP = triple superphosphate; GWP = global warming potential; EP = eutrophication potential; AP = acidification potential; NREU = non-renewable energy use; HTP = human toxicity potential; POCP = photochemical ozone creation potential; ECOTOX = ecotoxicity; ODP = ozone depletion potential

European production). In addition, data on ammonium sulfate production are available as well for Europe and Rest of the World.

7.3.2.3 PCR and EPD on Fertilizers

There is only one Product Category Rule (PCR) of fertilizers as defined by ISO 14025:2006, i.e. the PCR 2010:20 named "Mineral or chemical fertilizers" (EPD International 2016).

In this context, the main rules for preparing an Environmental Product Declaration (EPD) on fertilizers are briefly summarized. In detail, this document provides the PCR for the assessment of the environmental performance of UN CPC classes 3461, 3462, 3463, 3464 and 3465, according to the UN Central Product Classification system (UN CPC). The categorization of the UN CPC codes of interest is as follows:

Division: 34—Basic chemicals

- *Group: 346—Fertilizers and pesticides*

  - *Class 3461—Mineral or chemical fertilizers, nitrogenous*
  - *Class 3462—Mineral or chemical fertilizers, phosphatic*
  - *Class 3463—Mineral or chemical fertilizers, potassic*
  - *Class 3464—Mineral or chemical fertilizers containing at least two nutrients of nitrogen, phosphate and potash*
  - *Class 3465—Other fertilizers*

In Fig. 7.1, the scheme of the processes that have to be included in the above mentioned EPD is presented. The declared unit shall be defined as 1000 kg of product, including its packaging. The reference flow shall be defined at the customer gate, at the shelf or the retailer or at the market place. In the EPD a statement should be added to specify that the declared unit may have different functionality depending on the composition of the product that is declared. The processes are divided into three main groups: upstream, core, and downstream (Fig. 7.1).

Upstream processes (from cradle-to-gate) include:

- extraction of non-renewable resources (e.g. operation of oil platforms and pipelines);
- growing and harvesting of renewable resources (e.g. agricultural planting);
- refining, transfer and storage of extracted or harvested resources into feedstock for production;
- production processes of energy wares used in the extraction and refinement;
- manufacturing of materials and semi products;
- manufacturing of primary and secondary packaging.

Table 7.2 LCA studies not reviewed by Skowrońska and Filipek (2014)

References	Area of the study	Functional unit (FU)	Impact categories	System boundaries
da Silva and Kulay (2005)	Production of FMP and TSP (Brazil)	41,66 kg P_2O_5, which corresponds to 208 kg FMP (20% P_2O_5) and 93 kg TSP (45% P_2O_5)	GWP, ODP, HTP, freshwater aquatic ECOTOX, AP, EP; CML 2 baseline 2000	Cradle-to-gate
Hakala et al. (2012)	Organic and mineral fertilizers (Finland)	Unit biomass or energy	GHG emissions	Only the cultivation phase
Ridoutt et al. (2013)	Different P use efficiency in wheat growth (Australia)	1 kg of wheat	Abiotic resource depletion, freshwater eutrophication, GWP; CML, APSIM modelling, IPCC 2007	Cradle-to-field
Wu et al. (2015)	P use efficiency (PUE) of the crop production–consumption system (China)			Cradle-to-grave
Chiew et al. (2015)	Use of digested food waste as fertilizer and use of chemical fertilizer (Sweden)	A fertilizer containing 1 kg plant-available nitrogen and 0.20 kg phosphorus after spreading on arable land	GWP, AP, EP; IPCC 2006, CML 2001	Cradle-to-grave
Hasler et al. (2015)	Production and use of different fertilizer product types (Germany)	300 kg of complex fertilizer (with different nutrient composition) per ha	GWP, terrestrial AP, freshwater EP, NREU, resource depletion; ReCiPe midpoint (H)	Cradle-to-field
Zhang et al. (2017)	Production of DAP and MAP (China)	1 ton DAP (P_2O_5 45%, P 20%, N 17%) and 1 ton MAP (P_2O_5 43%, P 19%, N 10%), indeed 450 kg P_2O_5 and 430 kg P_2O_5	18 impact categories; ReCiPe midpoint and normalized values	Cradle-to-gate
Wang et al. (2017)	Different chemical fertilizer types (China)	Unit of N, P_2O_5, or K_2O	GHG emissions; PAS 2050	Cradle-to-field

The core process (from gate-to-gate) includes:

- external transportation to the core processes;
- production processes;
- recycling of waste or secondary materials for use in production;
- storage;

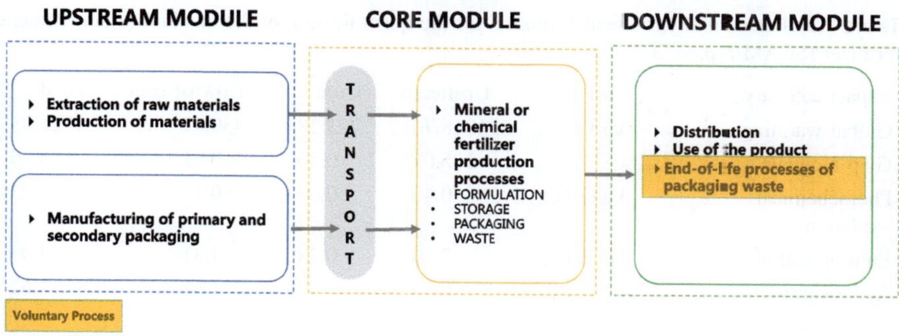

Fig. 7.1 System diagram illustrating the main processes and the division into upstream, core and downstream processes (EPD International 2016)

- waste treatment of waste generated during manufacturing;
- impacts due to electricity production according to the energy mix.

The downstream processes (from gate-to-grave) include:

- transportation from final manufacturing to an average retailer/distribution platform;
- the customer or consumer use of the product;
- end-of-life processes of packaging waste.

The potential environmental impact has to be reported for the following environmental impact categories, divided into core, upstream and downstream modules:

- emission of greenhouse gases (expressed as the sum of global warming potential, GWP, 100 years), in carbon dioxide (CO_2) equivalents;
- emission of acidifying gases (expressed as the sum of acidification potential, AP) in sulphur dioxide (SO_2) equivalents;
- emission of gases that contribute to the creation of ground-level ozone (expressed as the sum of ozone creating potential, POCP), in C_2H_4 (ethylene) equivalents;
- emission of substances to water contributing to oxygen depletion (expressed as the sum of eutrophication potential, EP), in phosphate (PO_{43}^-) equivalents.

An available Environmental Product Declaration (EPD) based on this PCR has been realized by SCAM S.p.A., and updated in March 2016 (SCAM S.p.A. 2016). In this specific case, the EPD concerns 27 organo-mineral fertilizers including 18 fertilizer formulations (reference year for data: 2014). In fact, some products can have the same formulation but different packaging, therefore, a different commercial name. In this case, the functional unit is the production and use of 1000 kg of

Table 7.3 Potential environmental impacts for the quantification of the medium organo mineral fertilizer (SCAM S.p.A. 2016)

Impact category	UM	Upstream	Core	Downstream	Total
Global warming	kg CO_{2eq}	1048.70	223.28	439.87	1711.85
Gwp biogenic		−185.07	0.50	−0.01	−184.59
Photochemical oxidation	kg C_2H_{4eq}	0.40	0.003	−0.40	0.03
Eutrophication	kg PO_4^{-} eq	2.40	0.23	1.31	3.93
Acidification	kg SO_{2eq}	8.83	1.17	1.61	11.62

packaged fertilizer and the system boundaries are divided into three standard modules:

- Upstream module that includes the production of the ingredients and the fertilizers packaging production;
- Core module that includes the manufacturing phase;
- Downstream module that includes products distribution and the use phase (emissions into air and water after fertilizers spreading).

In particular, for the quantification of the impacts coming from the use phase it is necessary to calculate the Agronomic Efficiency Index (A.E.I.) and the Uptake Index (U.I.) through field trials, as specified in ANNEX 1 of PCR 2010:20 (EPD International 2016).

The potential environmental impacts for the hypothetical organo-mineral fertilizer are reported in Table 7.3.

7.4 LCA of the Fertilizer Production from Exhausted Extinguishing Powders (EEPs)

7.4.1 Goal and Scope Definition

This LCA study has been used as a decision tool in the step of the "Engineering of the pilot plant", in order to optimize design process choices and minimize environmental impacts.

Therefore, the goal of the present study is to assess the environmental impacts associated with the innovative process for the treatment of EEPs aimed to produce fertilizers. The analysis follows the methodology defined by ISO 14040 and 14044 (ISO 2006a, b) and it is performed using SimaPro 8.3.

Functional unit and system boundaries

For this LCA study, the boundaries are set from cradle-to-gate so that the analysis starts from the arrival of the exhausted extinguishing powders in the factory and ends when the micro-fertilizers are produced in the factory. Considering

EEPs as a waste they do not carry any burden from the previous life, apart from those connected with the collection system. In this way, the innovative production process to be implemented, which is based on exhausted extinguishing powders at their end-of-life, that can be compared to the standard fertilizer production from virgin raw materials (traditional way). For the cradle-to-gate assessment the system considered is that presented in Fig. 7.3.

In this initial cradle-to-gate analysis, a declared unit (DU) is considered since the specific function of the product studied (fertilizer) is not known and this preliminary study does not include all the phases of the whole life cycle. Indeed, the concept of functional unit comprises function, quantity, duration, and quality of the analysed product, while the declared unit is used when the specific function of the analysed product is not yet defined or when the LCA does not cover the whole lifecycle (cradle-to-grave), but it ends at the gate of the factory (cradle-to-gate), as in the present case study.

The declared unit is defined as a flow of 1 ton of exhausted extinguishing powders (EEP) entering the ProPHOS plant. The declared unit could also be the output flow of fertilizer produced, but in this case, the choice of the input flow of EEP is done in accordance with typical LCAs of waste treatment.

Indeed, also in this case the zero burden hypothesis is assumed, i.e. waste (EEP in this specific case) does not carry any environmental burdens when delivered to the gate of the treatment plant: this choice is in line with the approach usually used in LCA of waste treatment (Ekvall et al. 2007).

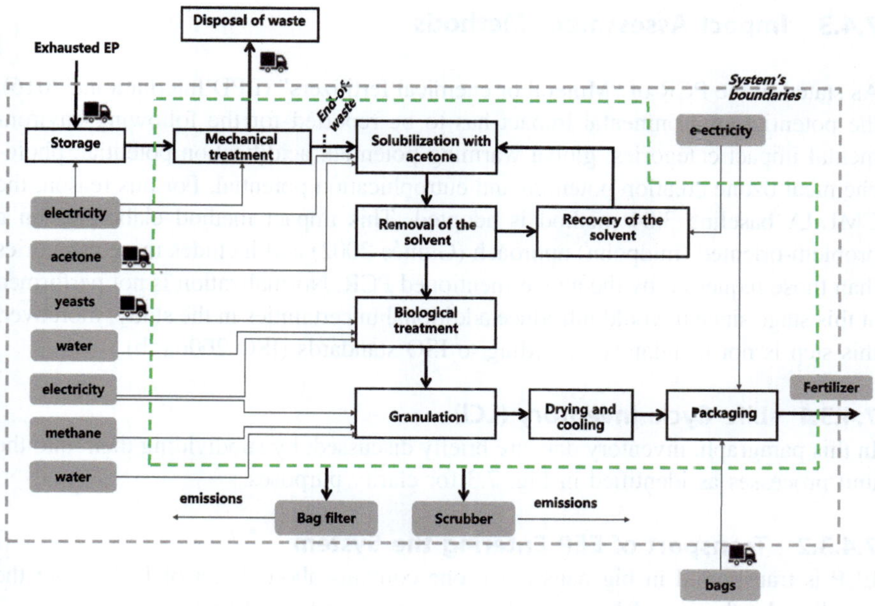

Fig. 7.2 System boundaries of the innovative recovery process; core processes are those highlighted by the green dotted line

In the next steps of the LCA (deliverables 2 and 3), the declared unit will be replaced by the functional unit, and it will be necessary to consider the function of the fertilizer itself, which depends on the nutrient content. The system boundaries and all the involved unit processes considered in this preliminary LCA are shown in Fig. 7.2.

7.4.2 Data Source, Quality and Allocation

In the present study, all the data regarding the process are primary data resulting from the small scale pilot project phase. Indeed, all the powers expressed are nominal data from the manufacturer's datasheet, therefore the consequent energy consumptions will be probably lower. At present, however, this fact cannot be checked before the industrial pilot-scale plant is operative. Also, data on the transport of EEP entering the plant are primary data obtained directly from the company, based on the average transport of EEP from the suppliers in 2016. The upstream processes of energy generation, materials production and transports are taken from the Ecoinvent v3.3 database. In detail, for electricity to be used in the plant the Italian mix is selected, while European conditions are considered for transports and production of fuels. For the moment, the downstream processes related to the disposal of wastes produced within the plant are not considered, as primary data are not yet available.

7.4.3 Impact Assessment Methods

As stated in the PCR of "Mineral or chemical fertilizers" (EPD International 2016), the potential environmental impact has to be reported for the following environmental impact categories: global warming potential, acidification potential, photochemical ozone creation potential and eutrophication potential. For this reason, the CML-IA baseline 3.04 method is adopted. This impact method elaborates on a problem-oriented (midpoint) approach (Guinée 2002) and includes more categories than those requested by the above-mentioned PCR. Normalization is not performed at this stage since it would introduce additional uncertainties in the study; moreover, this step is not mandatory according to ISO standards (ISO 2006a, b).

7.4.3.1 Life Cycle Inventory (LCI)

In this paragraph, inventory data are briefly discussed, by subdividing them into the unit processes as identified in Fig. 7.3 for clarity purposes.

7.4.3.2 Transport of EEP Entering the System

EEP is transported in big bags (each one contains about 1 ton of EEP) from the suppliers by the use of lorries with a maximum load of 14 tons.

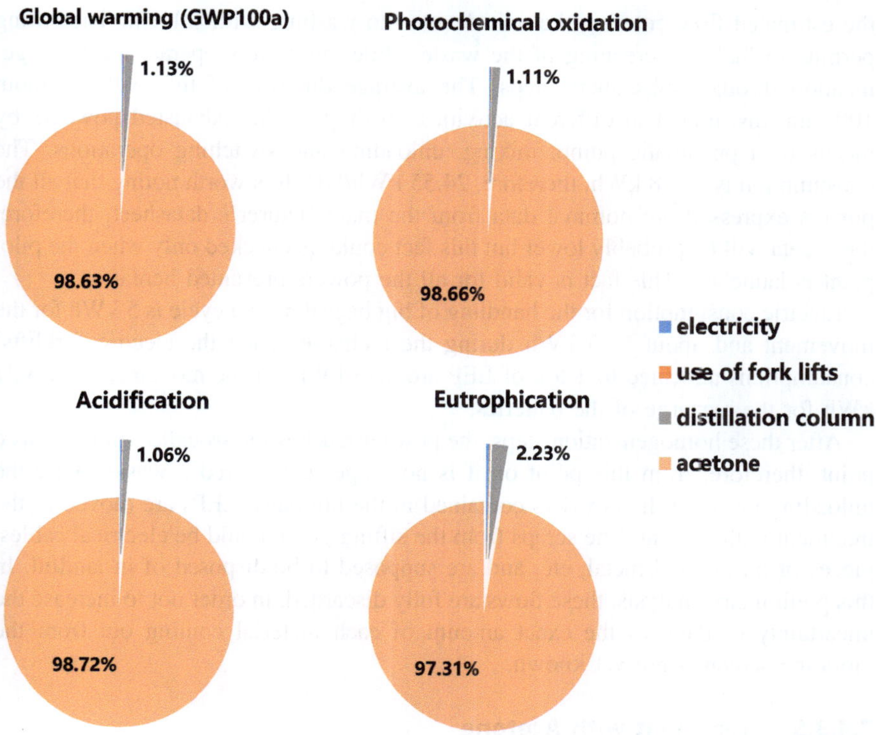

Fig. 7.3 Percentage contribution to the four main impact categories from different activities related to the unit process "treatment with acetone" in the scenario of no acetone reuse

7.4.3.3 Storage of EEP

The exhausted extinguishing powders (CER 160509) are stored in a dedicated area of 202.5 m^2. Next to this area, there is an equal area that is actually used to store used extinguishers (CER 160505) and an area of 207 m^2 that is used by the operators.

The maximum stored quantity of EEP per year is 2000 tons. Assuming that the area is completely occupied by the EEP, therefore, about 200 tons of EEP is present, the area needed for each ton of EEP is equal to 1,013 m^2. In the storage only electricity for lighting is needed, this result equal to 0.338 kWh for the area dedicated to the powders' storage. Machineries for handling the waste in big bag are all electric forklifts that are able to transport 1 ton of EEP. Electric consumption allocated to 1 ton of handled EEP is 0.50 kWh for the movement and about 0.51 kWh during the recharge of the batteries.

7.4.3.4 Mechanical Treatment

After the storage, the EEPs are sifted with a vibrating screen and then loaded into two mixers. A complete cycle includes the treatment of 1600 kg of EEPs as this is the estimated flow entering the unit devoted to washing with solvent. The sifting permits to make a screening of the waste while the mixers operate the homogenization through subsequent steps. The average duration of the cycle is about 100 min, distributed in different activities: loading of the exhausted powders by means of a pneumatic pump, mixing, unloading and switching operations. The consumption is 39.28 kWh, therefore, 24.55 kWh/DU. It is worth noting that all the powers expressed are nominal data from the manufacturer's datasheet, therefore, these data will be probably lower but this fact could be checked only when the pilot plant is launched. This fact is valid for all the powers presented hereafter.

Electric consumption for the handling of big bags during a cycle is 5 kWh for the movement and about 5.13 kWh during the recharge. Then the electric forklifts' consumptions allocated to 1 ton of EEP are 3.13 kWh for the movement and 3.21 kWh for the recharge of the batteries.

After these homogenization steps, the powder reaches the so-called end-of-waste point, therefore, from this point on it is no longer considered a waste. After the unloading phase of the powders contained in the big bags, EEPs are moved to the treatment with solvent. The scraps from the sifting phase could be electrical cables, pieces of plastic and metal, etc. and are supposed to be disposed of in landfill. In this preliminary analysis, these flows are fully discarded, in order not to increase the uncertainty of data, as the exact amount of each material coming out from the vibrating screen is not yet known.

7.4.3.5 Treatment with Acetone

The exhausted extinguishing powders are then loaded in the batch reactor together with an aprotic solvent (acetone) in order to enable the organic components to be removed and to produce a suspension. This phase is composed of five washing steps with acetone, carried out under stirring in order to further promote and accelerate solubilisation of the organic components.

The weight ratio between solvent and powders is 1:1 in the first step and then the ratio of acetone decreases; for the whole cycle (treatment of 1600 kg of EEPs in 2 h) this ratio is between 3.4 and 5. In this design phase LCA, an average ratio of 4.2 is considered; therefore, the amount of acetone is estimated to be equal to 6720 kg. Electricity consumption for the stirring operation is estimated being equal to 37.50 kWh/ton EEP. Electric consumption for the forklifts is the same as in the mechanical treatment because the big bags movements for a complete cycle require 20 min.

The global yield of the process (quantity of recovered raw material) reaches at least 98.5%, as there is no loss of material. The only discernible difference regards the possible presence of moisture in the EEPs collected from the suppliers; for the protocol of acceptance, the maximum limit is set to 5% by weight, but on average, it does not exceed 1.5%.

The present phase includes also the distillation column able to recover the acetone. First estimates give a consumption equal to 50 kWh per ton of reprocessed acetone. At the end of the solubilisations with acetone, 99.9% of the silicone oil is removed. In order to remove the residual traces of silicone oil, it is necessary to perform a biological treatment with yeasts.

7.4.3.6 Biological Treatment

The average duration of the cycle is about 90 min, considering loading of the exhausted powders by means of a pneumatic pump, mixing, spraying of the yeasts, unloading and switching operations. The consumption is 37.11 kWh for the cycle of 90 min, therefore, 23.20 kWh/DU. The required amount of yeasts varies from 1 to 3% by weight of input powders. In this analysis, a value of 48 kg of yeasts per cycle is considered in order not to underestimate the impacts. The electric forklifts for handling the powders are required also in this phase; their activity for a complete cycle requires about 20 min.

7.4.3.7 Granulation

After the biological treatment, the obtained molasses are added to the granulator. Regarding the consumptions of this phase, they consist of 10 kWh/DU for the loading and mixing phase, and 137.50 kWh/DU for the granulation. Methane needed for the warming operation is equal to 50 m^3/DU, while water required is equal to a maximum of 10% w/w.

7.4.3.8 Packaging

After drying and cooling of the produced granular fertilizer, there is the packing phase that requires 2 h of work and 8 kWh/DU in total. The final product is then packed into big bags of 1 ton or bags of 25 kg; these are generally provided by the customer or are already available in the plant; therefore, their production is neglected for the moment. Anyway, if only bags are used, 40 bags are needed to pack 1 ton of final product.

In this LCA, the declared unit is to be considered as packed in 25 kg bags.

7.4.3.9 Treatment of Emissions to Air

Air emissions generated within the plant are treated with bag filters and a scrubber. For the self-cleaning bag filters, due to their long-lasting use, it is not reported the quantity of filters required as well as the amount of dust removed after their rehabilitation. The emissions from the bag filter are below the Italian legislative limits and are allocated to 1 ton of treated EEP. In addition to the dust emitted throughout the process, during the heating the product releases ammonia, therefore, the entire line over which the product passes is directly connected with the scrubber in order to remove ammonia emissions. Even during bagging, any dust emitted is fed through filters directly connected to the scrubber.

7.4.4 Life Cycle Impact Assessment (LCIA)

If the acetone is not reused in the process, more than 95% of the impacts of EEPs recovery process are due to the unit process of treatment with acetone. Clearly, such a high environmental burden is due to acetone production and transport. A simulation of this scenario is reported in Fig. 7.3.

Originally, this solution was taken into account to avoid the installation of a distillation column. In view of these preliminary results, the process design was immediately reconsidered.

A careful estimation of the loss of acetone in the distillation phase is needed in order to provide a reintroduction of fresh acetone at the beginning of every cycle. It is indeed made a preliminary assumption that only 2% w/w of acetone is lost during distillation. This choice introduces a safety factor compared to the limit required in the technical specifications for acetone loss that is 0.5% w/w.

Table 7.4 and Fig. 7.4 shows the impacts of the cradle-to-gate life cycle impact assessment, calculated allocating the burden of distillation as if the stored acetone could be used for a year (i.e. 800 treatment cycles). Indeed, in order to treat all the EEP entering the plant in a year are necessary about 800 treatment cycles, assuming that 1.6 tons of EEP is managed each cycle.

Although in the final scenario acetone is recovered, the largest impacts come from the acetone unit. This is mainly due to the consumption of electricity and methane, considering that a distillation column has to be included in the process. The remaining contributions with a certain relevance in terms of environmental impacts are the biological treatment, the transports from the EEP supplier, and finally the emissions from the scrubber that mainly affect the acidification and eutrophication potentials, i.e. AP and EP, due to the presence of ammonia in the exhausted fumes.

Discussion The LCA run in the process design phase has been used as a decision tool for the process choices. Indeed, it has highlighted the relevance of certain process parameters such as rate of acetone recovery after any treatment cycle and the purity of recovered acetone. Moreover, possible recycling of acetone within a cycle (among the five washing steps) has to be evaluated, as it could involve a lower amount of acetone required (ratio acetone/powders lower than 3.4 by weight).

To test the possibility of further reducing the impacts, a second solvent has been considered, i.e. methyl-ethyl ketone (MEK). Among those possible, the choice of MEK seems a viable option from the point of view of environmental performances, having its production lower impacts in three out of four impact categories (Fig. 7.5). However, considering all the impact categories of the CML method, the results are no longer so definitive (Fig. 7.6).

As expected, the lower amount of MEK per DU has a sensible influence on the outcomes (Fig. 7.7); indeed, in most categories MEK process outperforms acetone one, e.g. compare results of Table 7.4 and 7.5.

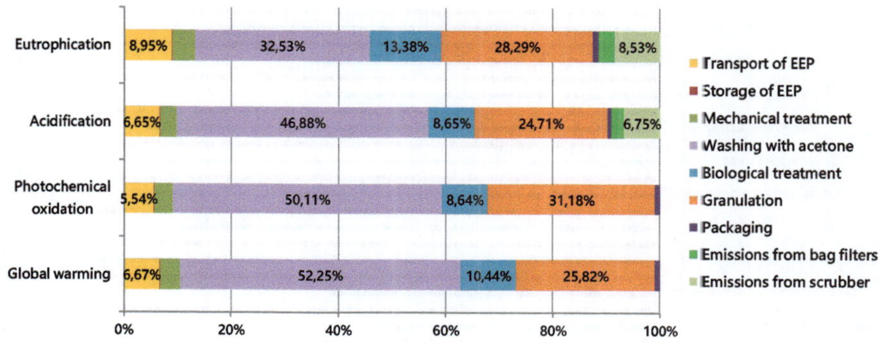

Fig. 7.4 Percentage contribution to the four main impact categories from different activities considering the reuse of acetone

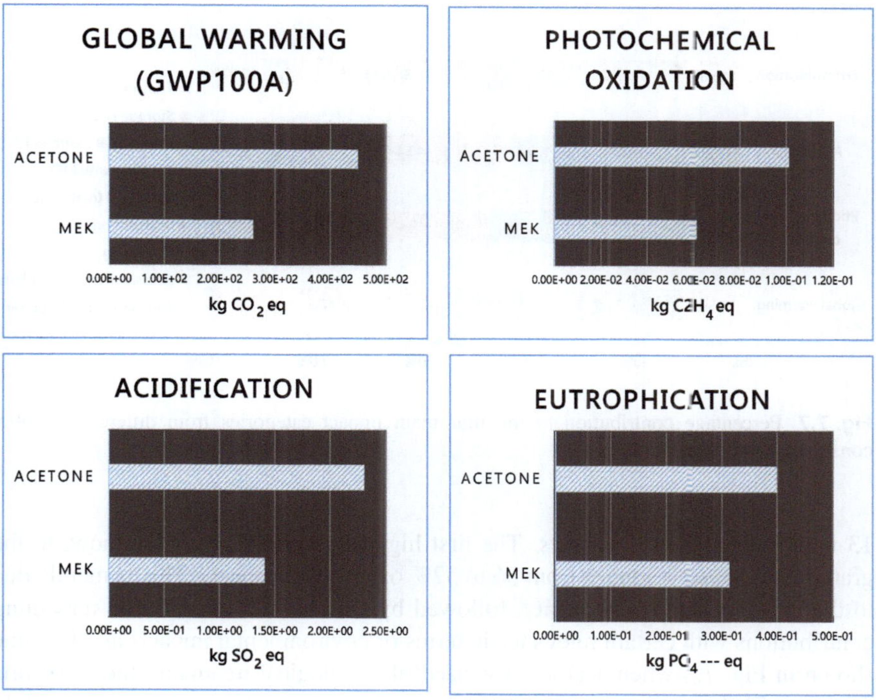

Fig. 7.5 Impact assessment results for the treatment of 1 ton EEP, using the two different solvents

In Table 7.5 is reported the value of total impacts associated with the MEK process for the four main categories of impact. Figure 7.7 shows that the treatment with MEK have a smaller contribution to the life cycle: its contribution ranges from

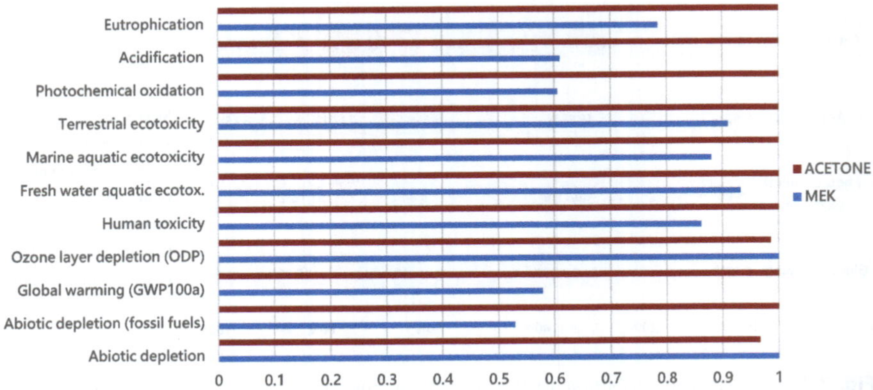

Fig. 7.6 Impact assessment results for the treatment of 1 ton EEP, using the two different solvents (MEK in blue and acetone in red)

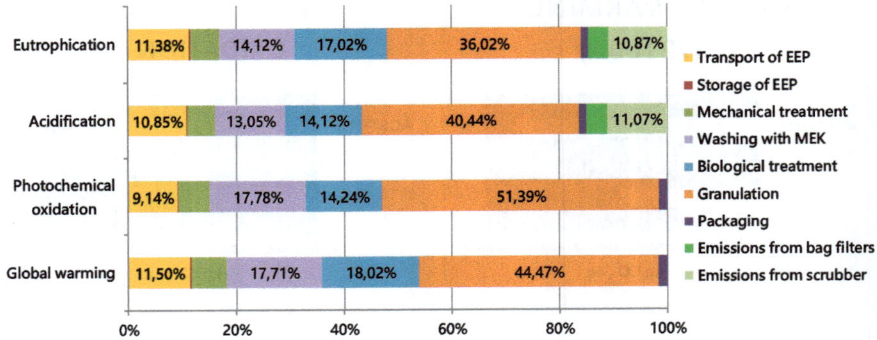

Fig. 7.7 Percentage contribution to the four main impact categories from different activities considering the reuse of MEK

13 to 17% of the total impacts. The first highest contribution is due, now, to the granulation phase: it ranges from 36 to 52% of the total impacts. This is mainly due to the consumption of electricity, followed by the use of methane. The remaining contributions with certain relevance in terms of environmental impacts are the same shown in Fig. 7.4 when acetone was used: the biological treatment, the transports from the EEP supplier, and finally the emissions from the scrubber that mainly affect the acidification and eutrophication potentials, i.e. AP and EP, due to the presence of ammonia in the exhausted fumes.

Table 7.4 Total impacts associated to the process from cradle-to-gate considering reuse of acetone

Impact category	Unit	Total	Transport of EEP	Storage of EEP	Mechanical treatment	Washing with acetone
GWP	kg CO_2 eq	4.50E+02	3.00E+01	5.51E-01	1.68E+01	2.35E+02
POCP	kg C_2H_4 eq	1.01E-01	5.60E-03	1.16E-04	3.54E-03	5.06E-02
AP	kg SO_2 eq	2.31E+00	1.53E-01	2.39E-03	7.27E-02	1.08E+00
EP	kg PO_4^{3-} eq	3.99E-01	3.57E-02	5.51E-04	1.68E-02	1.30E-01
Impact category	Unit	Biological treatment	Granulation	Packaging	Emissions from bag filters	Emissions from scrubber
GWP	kg CO_2 eq	4.70E+01	1.16E+02	4.34E+00	0.00E+00	0.00E+00
POCP	kg C_2H_4 eq	8.73E-03	3.15E-02	9.17E-04	0.00E+00	0.00E+00
AP	kg SO_2 eq	1.99E-01	5.70E-01	1.88E-02	5.28E-02	1.56E-01
EP	kg PO_4^{3-} eq	5.34E-02	1.13E-01	4.35E-03	1.16E-02	3.41E-02

Table 7.5 Total impacts associated with the process from cradle-to-gate considering reuse of MEK

Impact category	Unit	Total	Transport of EEP	Storage of EEP	Mechanical treatment	Washing with MEK
GWP	kg CO_2 eq	2.61E+02	3.00E+01	5.51E-01	1.68E+01	4,62E+01
POCP	kg C_2H_4 eq	6.13E-02	5.60E-03	1.16E-04	3.54E-03	1.09E-02
AP	kg SO_2 eq	1.41E+00	1.53E-01	2.39E-03	7.27E-02	1,84E-01
EP	kg PO_4^{3-} eq	3.14E-01	3.57E-02	5.51E-04	1.68E-02	4.43E-02
Impact category	Unit	Biological treatment	Granulation	Packaging	Emissions from bag filters	Emissions from scrubber
GWP	kg CO_2 eq	4.70E+01	1.16E+02	4.34E+0	0.00E+00	0.00E+00
POCP	kg C_2H_4 eq	8.73E-03	3.15E-02	9.17E-04	0.00E+00	0.00E+00
AP	kg SO_2 eq	1.99E-01	5.70E-01	1.88E-02	5.28E-02	1.56E-01
EP	kg PO_4^{3-} eq	5.34E-02	1.13E-01	4.35E-03	1.16E-02	3.41E-02

Conclusions

The recovery process of EEPs could produce fertilizers from a waste that now is disposed of with a non-negligible environmental cost. Although the process could be certainly improved, a comparison between the performances of fertilizers produced by primary raw materials and the present recovered MAP, leave no doubts about the virtuosity of the circular economy approach. Moreover, in terms of district symbiosis both industry sectors, i.e. extinguishing powders production and fertilizer production, can greatly benefit from a similar interchange of waste to secondary raw materials.

References

Ahlgren S, Bernesson S, Nordberg Å et al (2010) Nitrogen fertiliser production based on biogas—energy input, environmental impact and land use. Biores Technol 101:7181–7184
Brentrup F, Küsters J, Kuhlmann H et al (2001) Application of the Life Cycle Assessment methodology to agricultural production: an example of sugar beet production with different forms of nitrogen fertilisers. Eur J Agronom 14:221–233
Brentrup F, Küsters J, Lammel J et al (2004) Environmental impact assessment of agricultural production systems using the Life Cycle Assessment (LCA) methodology II. The application to N fertilizer use in winter wheat production systems. Eur J Agronom 20:265–279
Charles R, Jolliet O, Gaillard G et al (2006) Environmental analysis of intensity level in wheat crop production using life cycle assessment. Agric Ecosyst Environ 113:216–225
Chiew YL, Spångberg J, Baky A et al (2015) Environmental impact of recycling digested food waste as a fertilizer in agriculture—a case study. Resourc Conserv Recycl 95:1–14
da Silva GA, Kulay LA (2005) Environmental performance comparison of wet and thermal routes for phosphate fertilizer production using LCA—a Brazilian experience. J Clean Prod 13:1321–1325
Ekvall T, Assefa G, Björklund A et al (2007) What life-cycle assessment does and does not do in assessments of waste management. Waste Manag 27:989–996
EPD International (2016) PCR 2010:20. Mineral or chemical fertilizers. UN CPC classes 3461, 3462, 3463, 3464 & 3465. Version 2.0, date 2016-01-11. International EPD® System
ESPP (2017) European sustainable phosphorus platform. http://www.phosphorusplatform.eu/
EU (2019) Official Journal of the European Union, L 170, vol. 62, 25 June 2019
Guinée J (2002) Handbook on life cycle assessment—operational guide to the ISO standards. Int J Life Cycle Assess 6:255
Hakala K, Nikunen HM, Sinkko T et al (2012) Yields and greenhouse gas emissions of cultivation of red clover-grass leys as assessed by LCA when fertilised with organic or mineral fertilisers. Biomass and Bioener 46:111–124
Hasler K, Bröring S, Omta SWF (2015) Life cycle assessment (LCA) of different fertilizer product types. Eur J Agrono 69:41–51
Heffer P, Prud'homme M (2016) Fertilizer outlook 2016–2020. In: 84th IFA annual conference. International Fertilizer Industry Association (IFA), Moscow (Russia)
ISO (2006a) EN ISO 14040:2006—Environmental management—Life cycle assessment—Principles and framework. International Organization for Standardization
ISO (2006b) EN ISO 14044:2006—Environmental management—Life Cycle Assessment—Requirements and guidelines. International Organization for Standardization
ISO (2006c) ISO 14025:2006 - Environmental labels and declarations—type III environmental declarations—principles and procedures. International Organization for Standardization
Johnson EP, Clarke EK. Banks RE et al. (1997) Fire extinguishers: a case study of CFC replacements. Int J Life Cycle Assess 2:135–140
Linderholm K, Tillman AM, Mattsson JE (2012) Life cycle assessment of phosphorus alternatives for Swedish agriculture. Resour Conserv Recycl 66:27–39
Nemecek T, Huguenin-Elie O, Dubois D et al (2011) Life cycle assessment of Swiss farming systems: II. Extensive and intensive production. Agric Syst 104:233–245
PotashCorp (2014) Global fertilizer trends. http://www.potashcorp.com
Praticò FG, Moro A, Ammendola R (2010) Potential of fire extinguisher powder as a filler in bituminous mixes. J Hazard Mater 173:605–613
Ridoutt BG, Wang E, Sanguansri P et al (2013) Life cycle assessment of phosphorus use efficient wheat grown in Australia. Agric Syst 120:2–9
SCAM S.p.A (2016) Dichiarazione ambientale di prodotto di fertilizzanti organo minerali granulari. International EPD® System

Simonson M, Andersson P, Blomqvist P et al (2005) Environmental assessment of fires in products using the Fire-LCA model. In: Eighth international symposium of fire safety science. International association for fire safety science, pp 1071–1082

Skowrońska M, Filipek T (2014) Life cycle assessment of fertilizers: a review. Intern Agrophys 28:101–110

Tuomisto HL, Hodge ID, Riordan P (2012) Comparing global warming potential, energy use and land use of organic, conventional and integrated winter wheat production. Ann Appl Biol 161:116–126

United Nations Statistics Division (2017) UN CPC website—Central product classification, Ver. 2. https://unstats.un.org/unsd/cr/registry/regcst.asp?Cl=25

Zb Wang, Chen J, Sc Mao et al (2017) Comparison of greenhouse gas emissions of chemical fertilizer types in China's crop production. J Clean Prod 141:1267–1274

Wu H, Yuan Z, Gao L et al (2015) Life-cycle phosphorus management of the crop production–consumption system in China, 1980–2012. Sci Total Environ 502:706–721

Zhang F, Wang Q, Hong J et al (2017) Life cycle assessment of diammonium- and mono-ammonium-phosphate fertilizer production in China. J Clean Prod 141:1087–1094

MIX
Papier aus verantwortungsvollen Quellen
Paper from responsible sources
FSC® C105338

If you have any concerns about our products,
you can contact us on
ProductSafety@springernature.com

In case Publisher is established outside the EU,
the EU authorized representative is:
Springer Nature Customer Service Center GmbH
Europaplatz 3, 69115 Heidelberg, Germany

Printed by Libri Plureos GmbH
in Hamburg, Germany